Work, Politics and the Green Industrial Revolution

Douglas W. S. Renwick

Work, Politics and the Green Industrial Revolution

A Reflective Analysis of the UK Green Jobs Taskforce

palgrave
macmillan

Douglas W. S. Renwick
Nottingham Business School
Nottingham Trent University
Nottingham, UK

ISBN 978-3-031-53066-1 ISBN 978-3-031-53067-8 (eBook)
https://doi.org/10.1007/978-3-031-53067-8

This Palgrave Macmillan imprint is published by the registered company Springer Nature Switzerland AG.
The registered company address is: Gewerbestrasse 11, 6330 Cham, Switzerland

Paper in this product is recyclable.

To my parents with much love.

Acknowledgements

Thanks to Alec Selwyn and all at Palgrave Macmillan for the opportunity provided.

Contents

About the Author

Dr. Douglas W.S. Renwick is Associate Professor of Sustainable Work at Nottingham Business School, UK where he researches sustainable work. He has published over 30 articles in leading peer-reviewed journals, and an edited research text on green HRM by Routledge in 2020. Dr. Renwick is an Editorial Board Member of *Employee Relations: The International Journal, The German Journal of HRM, The International Journal of HRM,* HRM Section Editor for *Sustainability Letters* (Maastricht, EU), and is included in the Top 2% of Global Scientists. Doug is a senior research fellow at Rennes School of Business, France, and Director of Samsara HR Consulting, a non-profit sustainability consultancy. Address: Room 705, HRM Department, Nottingham Business School, Nottingham Trent University, UK, NG1 4FQ. Work telephone: + 44 (0)115 848 6429 Work email: douglas.renwick@ntu.ac.uk

1

Introduction and International Context

Abstract This opening chapter provides contextual background to this study. In it, I detail and discuss relevant global climate events, and how they link to politics and green jobs. Examples are drawn to support such points from the international context, the US, Donald Trump, Paris accord and Joe Biden eras. I also add more context from the European Union (EU), including EU green policies and criticism of them, plus EU green job creation. Such context frames the study herein.

Keywords Introduction • Context • US • EU • Politics • Green • Jobs • Creation

Global Climate Events, Politics and Green Jobs

The international context in HRM studies within which this research monograph occurs (Farndale et al., 2023) is the climate emergency brought on by global warming. Leading experts such as United Nations (UN) Executive Secretary Patricia Espinosa comment that greenhouse gas emissions have risen nearly every year since the original United Nations (UN) 1992 Paris Agreement and that the climate crisis is the

D. W. S. Renwick, *Work, Politics and the Green Industrial Revolution*,
https://doi.org/10.1007/978-3-031-53067-8_1

biggest threat to humankind (in Helm & Harvey, 2022, p. 3). Here, two processes occur: in governments producing commitments to limit greenhouse gas production, and in contradiction, them also extracting large fossil fuel reserves which produce potential tipping points to 'trigger cascading collapse' (Monbiot, 2022a, p. 3). In the United Kingdom (or Great Britain), which is the local country context herein, global warming is reflected in 21°C occurring just before Halloween in 2022 (Evans, 2022a, b, p. 1), the hottest British day seen 'since the summer of 1976' recorded in 2017 (BBC News, 2017a, p. 1), and the UK seeing a record temperature of 38.7°C (102F) in Cambridge on 25 July 2019 (AccuWeather, 2022, p. 3).

A specific impact arising from climate change is rising levels of pollution. Examples here include toxic waste from metal-mining pollution which impacts 23 million people globally (Gill, 2023, p. 1), and plastic pollution, where the journal *Nature Geoscience* reported one Manchester River has the most plastic pollution globally (Daily Express, 2018, pp. 1–3), and air pollution, where the UK High Court ordered the publication of draft government plans to cut air pollution and help safeguard public health, and the British Liberal Democrat politician Ed Davey argues that Ministers are using taxpayers money "to try to hide proof of their environmental failures" (in Laville, 2017, pp. 1–3). A relevant UK Conservative and Unionist Party (their official public name), or British 'Tory' party (their informal name; see Heywood, 2021, p. 71), government press release said levels of nitrogen dioxide (NO_2), mainly emitted by diesel vehicles, have been above legal limits since 2010 in 90% of urban areas, with relevant fumes causing approximately 23,500 early UK deaths every year. Indeed, the UK Treasury led by conservative George Osborne blocked 16 proposed clean air zones (CAZs), due to concerns about political fall-out and recommendations encouraging the purchase of low-pollution vehicles (Carrington, 2017, p. 3).

Here, the UK Labour Party argues that approximately 40 million British citizens live in areas with illegal levels of Nitrogen Dioxide (NO_2), with outdoor air pollution contributing to 40,000 early deaths a year, as NO_2 comes from UK factories and diesel engines, with NO_2 limits breached in 16 areas including London, Birmingham, Leeds and Glasgow (BBC News, 2017b, p. 3). Tragically, UK citizen Ella Kissi-Debrah aged nine from South London who had asthma became the first person globally whose death certificate formally stated air pollution as the cause of death (Ingham, 2021, p. 3). Related climate change impacts for the UK now include events where 'the numbers of climate refugees will dwarf anything we have seen, food imports will be problematic, and inland and coastal flooding will pose severe problems' (Hoskins, 2023, p. 15).

As such, part of the earlier history of UK conservative politicians' involvement in tackling climate change is, unfortunately, less community-minded, not reducing British air pollution and not always stimulating new, eco-employment or green jobs in increasing low-pollution transport. Thus, we can see connections between the climate crisis, British politics, apparent harm to UK citizens and few green jobs emerging above. Such links provide a foreground to concerns about how UK conservative government policies and actions impact UK green job creation—especially via their new Green Jobs Taskforce (GJT)—and any (un)intended consequences for all relevant stakeholders arising. As one author states, 'how politics and the environment mix [is] the framework within which policy on net zero and climate change is necessarily fixed' (Goodman, 2023, p. 21).

Before showcasing the relevant international political developments arising regarding green employment, I note that issues like pollution, climate displacement, wildfires, heatwaves and floods are large, growing UK and global problems, where 'people feel unprotected' (Smith, 2023, p. 1). For instance, European heatwaves killed 61,000 people (Squires et al., 2023, p. 13), UK heatwave deaths placed London as a 'high mortality risk' location (Sky News, 2023, p. 1), and temperatures across Europe rose above 40°C (104°F) in 2023 (Gatten, 2023a, p. 3), while in the UK London's air pollution was briefly seen as worse than in Beijing, with the most polluted UK city being Glasgow (Scotland), followed by Scunthorpe and Leeds with London in sixth place (Kirk & Scott, 2017, pp. 2, 4). Further, outdoor air pollution has claimed nearly 40,000 UK lives a year, and indoor air pollution is estimated to have led to approximately 99,000 deaths in Europe (Carrington, 2016, pp. 1–2). Reasons for such figures include too much carbon dioxide entering the atmosphere from factories and vehicles (Bloomberg News, 2020, p. 6).

The global picture is that nearly 10 million people lose their lives and 20 million are displaced and made homeless every year due to climate change (Sky News Live, 2023), while specific countries feel the impacts of global warming differently in the form of more extreme weather. For example, in the US commentators argue that combatting climate change (Napolitano, 2017, p. 2) is crucial because drought has plagued California since 2012 (Simpson, 2016, p. 2), and US wildfire pollution saw an estimated 1200–3000 excess deaths occurring in California in one 40-day summer period, with major fires also seen in Oregon and Washington (Milam, 2020, p. 2). Additionally, mudslides in Montecito killed over 20 people (Allen, 2018, pp. 1–2), flights were cancelled in Phoenix as it was too hot for planes (BBC News, 2017c, p. 1), and such events were played out again all over the US in 2023. The United Nations (UN) estimates that 178 million hectares of forest were lost worldwide over the past three decades, which is a land mass bigger than India and is why 63 countries, subnational governments and private organizations have now committed to restoring 173 million hectares of such forests (United Nations News (UN News), 2020, pp. 1, 3).

Climate displacement (where civilian populations are forced to move location) is also impacting US citizens in public housing. Here, US coastal communities ponder how to help people move away from waterside locales due to housing being damaged or destroyed by hurricanes and storms because people forced to leave rarely return, as seen in New Orleans after Hurricane Katrina and Superstorm Sandy, plus $2.9 billion of property in Charleston is at risk as it is lower than three feet above sea level (Flavelle, 2016, pp. 1, 4, 5, 7). Additionally, climate change has produced extreme floods in China in 2020 which affected more than 70 million people (Bloomberg News, 2020, p. 6), and the mega-flood of 2022 in Pakistan with water the size of the whole UK covering land. The picture above escalated in 2022–2023, with heatwaves globally seeing temperatures of 46.8°C in Taroudant and 46.2°C in Marrakesh, Morocco, and 46.7°C in Ouargla, Algeria (Extreme Temps, 2022, p. 1), records broken in Lytton, Canada, of 42.2°C, a new national record of 49.5°C in Turkey, and 81°C recorded in Iran with the highest heat index ever and dew point of 36°C—which is a new world record (Hudema, 2023, p. 1). Essentially, the climate change threat has worsened, as:

> Planet Earth has seen the hottest month on record. People suffered wildfires in Spain, Greece, Italy, Portugal, Hawaii, Canada, America, Australia, and Kazakhstan; record temperatures hit China and America; Chile's bizarre new heat record is 37C (99F)—in the middle of winter. The sea is as hot as a bath in Florida; polar scientists are scared by the vast loss of Antarctic Sea ice … we must try to adapt—but also to cut emissions, however tough politically that might be. (Harrabin, 2023, p. 15)

Thus, experts estimate that global temperatures of 1.5°C above pre-industrial levels are more likely to occur (a 66% likelihood) than not by 2028, which raises the prospect of more 'tipping point' events like the West Antarctic Ice Sheet crumbling and may produce a huge sea rise (Gatten, 2023b, p. 11).

Hence, climate change and government action or inaction to tackle it partly contribute to or help reduce human suffering as evidenced in the examples above. Ergo, government green policies matter. Another key link to help understand this book is that politics, employee voice and green issues all seem to interact, and the key relationship proposed herein is between politics and green jobs, following a politics-regulation-business-job creation/loss logic or continuum. One US-based company example of this relationship occurred at Amazon, which threatened to fire employees who spoke out about Amazon's climate policies, and upon seeing them, over 1000 Amazon employees protested (Palmer, 2020, pp. 1, 4, 5).

I now outline the international context which frames this study. In doing so, I discuss developments in two large market-based, advanced democracies and global carbon pollution emitters in the US and EU, because developments in them provide some stimulus towards explaining the key global context within which UK green job creation occurs. Additionally, my focus on the Trump regime and his withdrawal from the Paris Climate Agreement and EU developments are both important contextual events to explain relevant, arguable and relative UK climate scepticism among some conservative (Tory) politicians—especially the new British Prime Minister (PM) Rishi Sunak. Ergo, I attempt to set the scene here for later chapters.

International Context

The US

The global prospects for international governments acting to help reduce climate impacts and increase green jobs did not start well when both developments emerged in the mid-2010s (cf. Poschen, 2015). One reason for this was the then-US President Donald Trump's choosing to exit from the Paris climate change accord (Reuters News Agency, 2017, p. 2), which led to protests, as industry chiefs, corporations and business groups distanced themselves from the White House (Daily Telegraph and Agence France-Presse, 2017, pp. 4–5). Here, green activists argue that such US absence made it harder for the other 190 countries to achieve their agreed UN climate goals, as the US is responsible for approximately 15% of global emissions of carbon and previously committed $3 billion to help other countries adapt to climate change (Crilly, 2017, p. 2). Indeed, Trump's actions were problematic for the US, as they had led on green issues before under the Environmental Protection Agency (EPA) (Johnson, 2015, p. 2) and because at the G20 many other world leaders declared the Paris climate agreement as 'irreversible' (Asthana and Wintour, 2017, p. 1).

Trump and Paris

Part of Trump's rationale for withdrawing from the Paris Accord, and US initiatives like Superfund as part of a 31% budget cut to the EPA (Baptiste, 2017, p. 1), were detailed by his supporters and lobbyists representing Peabody Energy (the largest listed US coal producer), who said that, 'the accord, without significant changes, would have substantially impacted the US economy, increased electricity costs and required the power sector to rely on less diverse and more intermittent energy'. Moreover, Trump added that Paris could cost the US up to 2.7 million jobs by 2025 and produce costs for American workers and taxpayers of "lower wages, shuttered factories and vastly diminished economic production" (in Crilly, 2017, p. 2). Such cuts were potentially seen to harm millions of US citizens living near pollutants, especially lower-income neighbourhoods and minority communities, as roughly 53 million people live near a Superfund site (Baptiste, 2017, p. 2). Indeed, Trump's 'America First Energy Plan' embraced shale oil, gas fracking and clean coal technology plus boosting America's coal industry 'to bring jobs and prosperity to millions of Americans' (White House statement during Trump's inauguration speech, in Allen, 2017, p. 1). Trump claimed that the US Keystone XL Pipeline would also generate 28,000 good construction jobs, but other reports suggest he exaggerated this figure by sevenfold, and he was incidentally quoted on Twitter/X as saying that global warming was invented by China (Spencer, 2017, p. 3). Previously, Trump had also signed an executive order to 'get rid of other "job-creating" regulations', and for the US government to pass responsibility for such job generation over to the private sector 'via his credit-led infrastructure plan' (Shah, 2017, p. 3).

Part of Trump's plan was to deconstruct US climate change regulations established under the Obama administration (including the US Clean Power Plan trying to reduce main coal-fired power plant emissions), as his strategy was to increase US mining jobs (Crilly, 2017, p. 4). In doing so, Trump produced a do-it-yourself (DIY)-type greening for US businesses and consumers to potentially undertake, as he declared: 'I see some connectivity between humans and climate change' (Trump, 2016), meaning the status quo of a fossil fuel-dominant US economy would remain (Sky News, 2020, pp. 1–2). US protectionism came from Trump's Treasury Secretary Steven Minchin, who said that the US would prioritize "balanced trade" over lower border tariffs and climate change (in McGoogan and Sherlock, 2017, pp. 1–2), while Trump's EPA administrator Scott Pruitt begin a legal action against 28 states to stop the President Obama-led Clean Power Plan, and sided with ExxonMobil in Massachusetts and New York to oppose claims they misled investors by covering up their knowledge of climate change (Milman, 2016, pp. 1–2).

Overall, Trump's green record seems low, as he looked to de-fund the United Nations (UN) Climate Agreement; advocated for the US to become the largest global energy producer and reduce energy links with the Oil and Petroleum Countries (OPEC) (BBC News, 2017c); told advisers before the 2016 US election 'tell your miners, I got their backs'; had the US Republican Presidential candidate Rick Perry saying that Arctic drilling for oil and gas is possible; presided over the US fossil fuel lobby becoming highly active; and made anti-wind farm statements about them blighting his proposed golf course at Aberdeen in Scotland, UK (Channel 4, 2017). Such developments did not appear to support the creation of many new, US-based green jobs under Trump, and his actions provided a controversial anti-narrative to doing so. For example, the pro-fossil fuel jobs created by Trump were branded by various US media sources then as 'good' jobs because they were some jobs, i.e. new US-based employment, and supposedly US-based jobs too—which all complimented Trump's let's 'make America great again' (MAGA) political leitmotif. Additionally, his supporters viewed such opportunities as 'green jobs', because they may help secure immediate and medium-term US energy supplies quickly and cheaply, meaning who could argue with Trump's allegedly US-based 'green job-creation' related outcomes? Most recently, Trump criticized Joe Biden's "ridiculous regulations and environmental extremism", which would "decimate" the US car industry, with a "maniacal push" on electric vehicles which "will take away US jobs and empower China" (in Coen, 2023, p. 12).

The Trump legacy of very few, new green jobs seems important to state, as fresh US Presidential elections are coming up in 2024, and Trump may be a Republican leadership candidate then. This is because when asked if his being sentenced under charges of mishandling US state secrets in 2023 would stop his campaign, he said: "Not at all. There's nothing in the constitution to say that it could" (Sabur, 2023, p. 14). Thus, a particular danger arising for the US with any potential future Trump administration is one where green job gains made under the new Joe Biden administration possibly vanish under Trump's pro-fossil fuel policies.

Trump and Biden

A big problem for Trump then, as it is now, is that the impacts of the climate crisis have worsened since he left office and because the politics of his own self-styled 'populism' have unravelled into his ultimate impeachment. Here, the US is now having to think ahead regarding climate change—not least to help protect their ecology—which is threatened by warming in Alaska, sea level rises on the East Coast, dust storms in Western Texas (Panhandle and South Plains) and Arizona, plus earthquakes in California and Yellowstone Park in 2022–2023. Notably, the US did not secure a more stable global order under Trump, i.e. with him as a Roosevelt-style President working with international partners to avoid an ecological catastrophe, and Trump's support for fracking is debatable because the short, medium and long-term impacts of it on US land sites are unknown. Overall, the US seems to have perhaps lost four years of environmental leadership under Trump, and he did not seem to be living in truth (cf. Havel, 1987) as US President, because his calculation around 'Trump digs coal' only seemed designed to protect US-based miners, and for his team and supporters 'to avoid regulation and taxation' (Monbiot, 2016, pp. 1–2). Moving forward, reporters note that 'Trump's actions and justification for quitting the Paris deal have been rejected' by the new US President Joe Biden, who has said he will rejoin the deal as soon as he takes office (Sky News, 2020, p. 4).

One emerging contradiction in the US is that while they support 'industry-based solutions' through creating '2.2 million American clean energy employees' (Chung, 2017, p. 2), the new Joe Biden administration faces three key barriers to enacting his own, new green jobs plan, namely: bureaucracy, permitting and a lack of regulatory mandates (The Economist, 2023a, p. 31).

Thus, the centrifugal forces of conservatism in post-Trump US politics remain strong and partly shape the regulatory environment within which new US green job creation occurs. Such forces may perhaps restrict progress in achieving Joe Biden's stated green job plans, meaning that it remains to be seen, even with the huge financial level and wide scope of Biden's eco-jobs grand plan (detailed below), how many new, decent, US-based green jobs can be created, due to such conservativism also occurring.

Biden

The election of Trump's successor Joe Biden as US President did not automatically herald a new golden age for US-based green job creation and growth, as critics pointed to the future direction of US climate strategy being 'cloudy' (Napolitano, 2017, p. 1). This is because Biden 'would not sign documents committing to phase out US coal power by 2030, making no legal committing to adopt net zero, only agreeing to make a non-binding pledge to reduce emissions by 50% and signing a climate bill which seems protectionist', as it allegedly offers 'bungs to buyers of US-made electric cars' (Clark, 2022, p. 2). Additionally, Biden has:

> A more open posture towards Big Oil. Witness his approval of Willow, an Alaskan oil project and the fine print of the BIL and IRA. The laws offer subsidies for decarbonization technologies well beyond wind and solar to include carbon capture and sequestration (CCS) and hydrogen (made from fossil fuels). (The Economist, 2023a, p. 31)

Nonetheless, upon taking office Biden showed his pro-green credentials in signing an executive order to rejoin the Paris climate agreement, rolling back the Keystone XL Pipeline and pledging $2 trillion to create millions of jobs in energy retrofits and clean energy (Holland, 2021. pp. 4, 5, 7). Thus, some commentators stress Biden's claim that "America is back" and "ready to lead" again (in Daily Telegraph, 2020, p. 1), as his administration has formalized local content policies to deliver factory investment (Birkett et al., 2022, p. 1). Further, from Biden's administration web page, he states that the US employs over 3 million people in the clean energy sector and that his strategic response to climate change 'can create more than 10 million well-paying jobs in the US' and enact new partnerships to develop programmes training the US workforce to gain access to clean energy jobs. Indeed, he adds that he will 'incorporate skills training into infrastructure investment planning, so that every American can participate', and plans for the US to be the world leader by 2030 'in clean energy research, investment, commercialization, manufacturing, and exports' (Biden, 2021, pp. 1–2).

Biden continues that his plan will 'create the industries of the future, by investing $400 billion over ten years to develop new technological breakthroughs', US companies will supply low-carbon solutions to international markets to help US industry, and he will 'implement a national strategy where states or regions can build low-carbon manufacturing via tax credits and subsidies' (Biden, 2021. pp. 2–4). Biden states that his plan 'will establish a Task Force on Coal and Power Plant Communities', to help them leverage federal and private sector investments, and generate 'high-paying union jobs' and partnerships with unions and community colleges 'to create training opportunities for these new jobs' (Biden, 2021, p. 1).

So far, green job progress under Biden includes energy technology leader Plug Power building a huge plant to make liquid hydrogen; clean investment trends above forecasts; and Biden's three recent laws, the Bipartisan Infrastructure Law (BIL), CHIPS Act and IRA, seeing $200 billion of firm investments into batteries, electric vehicles (EVs), renewables and hydrogen (The Economist, 2023a, p. 31). Here, the IRA's climate-focused provisions include '$369 billion in tax credits and direct government funding towards energy and infrastructure, and clean hydrogen ones not capped by Congress, meaning that if investors like them, IRA public climate spending might reach $800 billion. If private capital influence is added, "$1.6 trillion could be made in decarbonization investments over the next decade, with more tax credits for solar and wind energy later on", and the IRA is "unlikely to be overturned by a future Republican Congress, in part because Houston benefits"' (The Economist, 2023a, p. 32). Additionally, a report by advocacy group Climate Power found that more than 100,000 new jobs were gained in 31 states before the end of January 2023, and over 75 decarbonization projects worth $1 billion are being developed, because as Democrat Senator Tom Wyden (Oregon) explains: "the more you reduce carbon emissions, the more tax credits you earn" (The Economist, 2023a, p. 32).

Yet Biden's plan to re-shape the US economy via Congress passing three bills on 'infrastructure, semiconductor chips and greenery' plus providing $2 trillion to remake the US economy also contains limitations, such as US protectionism annoying EU and UK allies and subsidies generating inefficiencies too (The Economist, 2023b, p. 9). Here, alongside cost overruns, skilled labour shortages and Biden approving new shale licences in Texas, specific challenges include moving energy from sunny and windy locations to ones with the highest demand and bypassing bureaucracy, i.e. of up to five years to get a solar farm permit and seven years to get an onshore wind farm one (The Economist, 2023c, p. 33).

Further, Biden's role in US workplace greening is now contested, as in his first 60 days he was concerned about his legacy, delivering a Roosevelt-type New Deal (Sopel, 2021, p. 9), and he is controversial among US

Democrat politicians because former US President Obama warned 'not to underestimate Biden's ability to muck things up' (Shipman, 2020, p. 1). Previous legacy problems could also undermine Biden's green plans, including injustices surrounding Washington state suing the US government over toxic vapours at the Hanford Nuclear Reservation, where over 50 instances were recorded since 2014 of workers exposed to hazardous fumes including chemical pneumonitis (Johnson, 2015, p. 1).

Nevertheless, key stakeholders such as Bill Gates state that mitigating climate change needs government research and development, as "the climate problem has to be solved in the rich countries" (in Cahill, 2015, pp. 3, 6, 7), including the US, and initiatives such as an old coal mine becoming the largest new solar farm in Kentucky show that the false narrative of climate action being 'a net job loser' can be contested (Grover, 2018, p. 2), and new US-based evidence is that environmental regulation increases demand for higher-level analytical or technical skills too (Valero et al., 2021, p. 4). I now outline and critique the European green jobs plan.

The European Union (EU)

The EU has produced a far-reaching Green New Deal to invest '30% of their £1.6 trillion long-term budget on climate-related measures', alongside Germany eclipsing their energy record and generating 85% of electricity from renewables, Portugal advancing well on solar use and Spain moving rapidly to ending their dependence on fossil fuels (European Commission, 2023c, p. 1; Houser, 2017, p. 2). The green employment picture across Europe is one where green job share varies from 17% in Greece to 22% in Germany, and such workers 'tend to be older, fewer are female, more are highly skilled, and more are likely to be on permanent contracts' (Valero et al., 2021, p. 6). Here, country-level differences include workers in green jobs being likelier to have a university degree (in Belgium, Greece, Luxembourg, the Netherlands and Spain), and a similar link existing for those in enhanced skills jobs (in Denmark and Luxembourg) (Valero et al., 2021, p. 6). Overall, around 20% of jobs in 14 European economies could be characterized as directly and indirectly green if a broad occupation-level definition is used (Valero et al., 2021,

p. 1), and a huge 14.6 million direct and indirect jobs already exist 'in biodiversity protection and forest and national resource rehabilitation' (Singh, 2018, p. 2) in the EU too.

EU Green Policies

In 2021 the EU passed laws to arrest deforestation (Razzouk, 2021, p. 1), introduced targets to remove 5 million tons of CO_2 from the atmosphere by 2030 (Rathi, 2021, p. 1), and will invest over €290 million in 132 new nature and climate action projects (European Commission, 2021a, p. 1). The EU just moved past the halfway mark in their goal to close all of Europe's coal plants by 2030 (Europe Beyond Coal, 2021, p. 1), aim to plant 3 billion trees by 2030 (Happaerts, 2021, p. 1) and passed a new criminal law to enforce green policies too (European Commission, 2021b, p. 1).

In 2022 the EU moved further, as their Green Deal resolutions included cutting methane emissions to neutrality by 2050, reducing air, water and soil pollution by 2030, investing in green fuels and infrastructure to support the take-up of zero-emission vehicles (including more charging points for electric vehicles), investing in long-term energy storage, renovating up to 35 million buildings (to help create jobs in construction) and increasing their overall EU renewables target to 40% by 2030 (European Commission, 2022a, p. 1). Additionally, the EU is focusing on green skills in education (European Commission, 2022b, p. 1) and investing over €1 billion in energy infrastructure (European Commission, 2022c, p. 1) while the Italian government are paying their citizens 100% of the cost of insulating walls and windows and installing heat pumps and solar (Spurway, 2022, p. 1).

Here, analysts comment that while the EU pace looks too slow to achieve their aim of zero-emission electricity by 2040 (Van der Leun, 2022a, b, p. 1), EU countries invested $219 billion in the clean energy transition in 2021 and EU renewable energy outpaced fossil fuels in 2020 and became the main EU power source for the first time (Polman, 2022, p. 1), while Spain commissioned new wind farms to produce 100% renewable electricity by 2023 (Johnson and Johnson, 2022, p. 1). The

EU set a record wind generation in 2022 (Beirao, 2022, p. 1) and is a global battery centre with 20 gigafactories (European Commission, 2022c, p. 1) and Germany is set to have fossil-free electricity by 2035 (Ember Climate, 2022a, p. 1). Further, the EU Commissioner for Climate Action Frans Timmermans states: 'let's dash into renewables at lightning speed' (in Evans, 2022a, b, p. 1), as the EU is considering a higher renewable energy target for 2030 (Taylor, 2022a, b, p. 1), it has already agreed a 45% renewable energy target by 2030 (Connolly, 2022, p. 1), Slovakia confirmed it will close their Novaky coal plant in 2023 (Europe Beyond Coal, 2022, p. 1) and Poland has become the seventh largest maker of solar panels in Europe (Ember Climate, 2022b, p. 1).

Moreover, President Macron pledged to make France "the first great nation to exit gas, oil and coal" by building more nuclear reactors, expanding solar by tenfold and constructing 50 offshore wind farms by 2050 (Ember Climate, 2022d, p. 1), MEPs voted to end the sale of petrol and diesel cars (Murray, 2022, p. 1), and national plans put the EU on course for 63% renewable electricity by 2030 (Ember Climate, 2022c, p. 1). The EU plan to install 20 million heat pumps by 2026 and 60 million by 2030 (Ralston, 2022, p. 1), the European Commission threatened sanctions against trade partners failing to match their sustainability standards (Business Green, 2022, 2022, p. 1) and Frans Timmermans said the EU is halving chemical pesticide use and banning it from all schools, parks and playgrounds (Timmermans, 2022, p. 1). The Netherlands are aiming for heat pumps as their default by 2026 (Electrify Heat, 2022, p. 1), solar and wind are planned to provide 70–80% of European electricity by 2035 (Ember Climate, 2022c, p. 1) and Denmark has agreed to increase wind and solar threefold, stop gas use for heating by 2035, replace existing gas demand with biogas and introduce a carbon tax on top of the existing EU one—all by 2030 (Capion, 2022, p. 1).

Additionally, achieving clean power by 2035 could save the EU €500 billion to €1 trillion (Ember Climate, 2022e, p. 1), the EU legislated to back mandatory use of green jet fuel from 2025 (Carrington, 2022, p. 1), and their 2030 coal exit is still on track (Ember Climate, 2022j, p. 1). Serbia may close the hugely polluting Kostolac A coal power plant (Ember Climate, 2022h, p. 1), Austria has had a Just Transition Fund plan approved by the EU which will invest €136 million in their most carbon-intensive regions (Happaerts, 2022, p. 1) and Spain powered up Europe's largest solar power facility (Mathis, 2022, p. 1). Germany and Denmark are scoping a new $9 billion offshore wind power hub to power 4 million homes (Razzouk, 2022, p. 1), the EU will invest up to €300 billion to quicken their green transition (European Commission, 2022c, p. 1) and solar may be the largest European energy source by 2030 (World Economic Forum, 2022, p. 1). Denmark, Germany, Belgium and the Netherlands are all planning to build four artificial islands to increase offshore wind fourfold in the North Sea (equal to 30 nuclear reactors) (Hudema, 2022, p. 1), and the EU has reduced their dependence on Russian gas from 40% to 7% of all imports (Ukraine World, 2022, p. 1).

Lastly here, in 2023 EU emissions are still falling among a return to coal (Harvey, 2023, p. 1), wind and solar were the top electricity source for the first time in the EU over 2022 (Carbon Brief, 2022, p. 1) and 38 green steel projects are planned across Europe, compared to only one in the UK (ECIU, 2022, p. 1). The EU issued €6 billion in green bonds on the capital markets (Hahn, 2023, p. 1) and is increasing the number of publicly available electric recharging and hydrogen refuelling stations (ECIU, 2023, p. 1), as Europe aims to cut their emissions by at least 55% by 2030 (European Commission, 2023b, p. 1). The EU has allocated €250 billion for green growth investment (European Commission, 2023a, p. 1), EU renewables provided 40% of all European electricity over the 2022 winter period (Razzouk, 2023, p. 1), the EU share of greenhouse gases fell by one-third between 1990 and 2021 (Carbon Tracker, 2023, p. 1) and the EU passed a plan to preserve nature and tackle climate change through restoring particular species and habitats (Covering Climate, 2023, p. 1). France has just banned internal short-haul flights where train alternatives exist (BBC News, 2023a, 2023b, p. 1), and the EU has now even agreed a common charger for mobile phones, tablets and cameras (USB type C) (European Commission, 2023d, p. 1). Then, EU green plans and actions are extensive.

EU Criticism

Nonetheless, despite the many and varied pro-green EU policies above, much criticism of them also occurs. Here, Greta Thunberg and scientists have criticized the EU's Common Agricultural Policy (CAP), which accounts for the largest single part of the EU budget, for producing biodiversity loss and as "not in line with the Paris agreement or the 1.5C target" (Thunberg in O'Leary, 2021, pp. 2, 4). France views nuclear power as green while Germany do not (The Economist, 2021, p. 1), the EU backed 30 gas projects totalling €13 billion (Blake, 2021, p. 1) and a global group of 60 investors warned against using nuclear and gas in the new EU taxonomy for sustainable finance (CAN Europe, 2021, p. 1). Some 70% of EU soils are not classified as in good condition (European Commission, 2021, p. 1), the EU-built environment contains the biggest climate investment gap in any sector (Green Finance Institute, 2021, p. 1), the EU ignore methane emissions from fossil fuels and imported coal (Ember Climate, 2021c, p. 1) and EU governments are shipping their plastic waste to other countries and exposing overseas citizens to toxic chemicals (Rall, 2021, p. 1).

Moreover, the 27 EU countries spend around €55 billion each year subsidizing fossil fuels (Morgan, 2022, p. 1); Poland, Czechia and Bulgaria will become responsible for more than 95% of planned EU electricity generation from coal by 2030 (Ember Climate, 2021a, p. 1), and Germany is responsible for most of the coal burnt in Europe (Ember Climate, 2021b, p. 1). Poland produced one-third of EU electricity from coal in 2021 (Ember Climate, 2022a, p. 1), and the EU electricity mix is viewed as setting too slow a pace for zero emissions by 2040 (Van der Leun, 2022a, p. 1). EU emissions are declining at half the rate needed to achieve 1.5°C because clean power is not outpacing fossil fuels quickly enough (Ember Climate, 2022b, p. 1), and Greta Thunberg describes the EU taxonomy as something which "takes greenwashing to a completely new level" (Thunberg, 2022a, p. 1). Germany has said that nuclear energy contains dangers and rejects EU plans to label it as sustainable (Politico Europe, 2022, p. 1), the EU aim to describe 1.4 billion tons of fossil gas as 'green by law' (Hoepner, 2022, p. 1), and most new EU renewables power will supersede nuclear energy instead of coal/gas generation (Jones, 2022, p. 1). Investors are requesting to ban fossil gas from the EU green finance rulebook (Business Green, 2022, 2022, p. 1), and the Swedish government is clearing forests and building mines in Sapmi (Axelsson, 2022, p. 1).

Further, EU plans to quadruple solar and wind capacity do not match the sixfold growth needed (Ember Climate, 2022d, p. 1), many EU states have few detailed plans to reduce fossil gas capacity by 2035, while some plan new capacity (Ember Climate, 2022c, p. 1), and none of the EU 21 power utilities plan to deliver the power sector milestones needed for 1.5°C (Cremona, 2022, p. 1). The Finnish company @Fortum plan to run coal-fired plants after 2030 (Ember Climate, 2022c, p. 1), a 200-year-old Netherlands forest is to be cleared so NedCar can build electric cars for @Rivian (Brightwell, 2022, p. 1), 25 EU banks provided €55 billion in 2021 for energy firms to increase oil and gas production (Green Party, 2022, p. 1) and EU oil majors and big mining companies are not shutting down coal mines and oil wells (Carbon Tracker, 2022a, p. 1).

Legislation is needed to confirm coal-fired power plants in Germany are closed by 2030 (Ember Climate, 2022e, p. 1), and EU audits reveal more tax breaks for fossil fuels than renewables (Carbon Tracker, 2022b, p. 1). Alongside the US, Europe is named as being behind the most global ecological damage (International Resource Panel, 2022, p. 1), wind power laws in Poland stop it from opening half the capacity needed to deliver the EU 2030 green goals (Ember Climate, 2022j, p. 1) and air pollution levels among Southeast Europe are over the World Health Organization guidelines (State of Global Air, 2022, p. 1). EU farmland may be the world's largest reserve of microplastics due to sewage fertilizer (Harrabin, 2022, p. 1), and Poland's Belchatow coal plant was Europe's dirtiest for emissions in 2021, which is its 17th time in first place (Ember Climate, 2022h, p. 1). EU plans to escape Russian fossil fuels include a 41% increase in coal use (Evans, 2022a, 2022b, p. 1) and an EU schedule to prioritize cash and permits for 30 gas projects faced a legal challenge from green groups due to it not considering methane emissions (Boffey, 2022, p. 1).

Additionally, 18 coal plants in Southeast Europe are estimated to be responsible for 19,000 deaths (Ainger, 2021, p. 2), and the EU have 'not even planted 1%' of their 3 billion tree-planting goals (Thunberg, 2022b, p. 1). The EU strategy to include green principles in infrastructure development is still underdeveloped (Chatham House, 2022, p. 1), one-tenth of flights leaving France in 2019 were private aircraft which generated equal CO_2 to one average EU citizen emitted all year long (Wallace-Wells, 2022, p. 1), and the EU Parliament voted to label fossil gas as 'green' energy (Thunberg, 2022c, p. 1). Higher EU renewables targets have not yet been deployed, and only four European countries are on course for wind power deployment in line with 1.5°C (Ember Climate, 2022l, p. 1), while bottom traw fishing is practised in 59% of EU marine-protected areas, releasing as much CO_2 as the aviation industry (McVeigh, 2022, p. 2). German wind power generation has slowed (Centre for Research on Energy and Clean Air, 2022, p. 1), some EU states want to lower the 45% renewables by 2030 target (Taylor, 2022a, 2022b, p. 1), and a draft of the EU net zero industry act includes nuclear as a strategic net zero technology (Genicot, 2023, p. 1). Europe needs upskilling and reskilling in energy-intensive sectors (Industrial Europe, 2023, p. 1), and the EU has far-reaching methane emissions from oil and gas plants (Clean Air Task Force, 2023, p. 1). The EU needs to invest an extra €700 billion per year to enact a green economy and exclude cheap Russian fossil fuels (Bloomberg Green, 2023, p. 1) and coal plant emissions pervaded the EU power sector in 2022, with Germany and Poland as the dirtiest emitters (Ember Climate, 2023, p. 1). The EU has also been criticized for not making environmental protection guarantees because of possible US tariff threats emerging; intensive farming methods in their reformed CAP and potential 'climate colonialism' due to the ecological damage European-based companies cause to the ecologies of the Global South countries too (see Baldry & Hyman, 2022, 2022, p. 86). Some observers even claim the German government 'may be toppled in 2023 by a mass revolt against heat pumps' (Warner, 2023, 2023, p. 21).

EU Green Jobs

The picture above means that more EU eco-employment looks possible. This is because the EU is keen to enhance building insulation, increase the speed of heat pump installations, set aside €113 billion for new hydrogen infrastructure, propose new legislation to make solar and wind farm builds easier, make solar panel installs on some roofs mandatory and enable REPowerEU legislation to facilitate renewable solutions (Fisher, 2022, pp. 2–4). The EU is including nine green hydrogen projects in Germany, investing €8.5 billion in decarbonizing steel, which the EU topped up by €1 billion in 2022 for green hydrogen at Germany's Salzgitter plant (ECIU, 2023, p. 1). Finland has installed heat pumps in one-third of its homes, the Netherlands are proposing to disconnect all their homes from the gas grid and Estonia offers free public transport (Monbiot, 2022b, pp. 3–4). Heat pump sales went up 40% in the EU to 3 million units in 2022 (Rolt, 2022, p. 1), and the EU voted to ask each member state to develop two cross-border projects to expand green electricity (European Parliament, 2022, p. 2). Additionally, the EU plans to fast-track and simplify permits needed for climate tech sites, temporarily loosen state aid rules in the short term and gain more from existing trade agreements (Howarth, 2023, pp. 1–3).

The EU has reached 25% of global wind employment (IRENA, 2023, p. 1) and plans more green jobs in solar too, as solar panels now generate more electricity across the EU than all the bloc's coal plants (Mathis, 2023, p. 2). Lithuania has completed €1 billion worth of energy-efficient building upgrades, which the EU wishes to scale up, as in Lithuania alone this project has created about 70,000 new jobs in green construction and the service sector (Capps, 2022, pp. 1, 10). The EU wish to keep jobs generated from using low-carbon technologies such as solar panels and heat pumps within the EU by making 40% of such tech made within EU borders by 2030 (Carbon Brief, 2023, p. 1). All of these initiatives above may provide new green jobs across the EU.

However, reductions in EU-based green employment also exist, as investment in new EU wind farms dropped to their lowest level in over a decade (Reierson & Ainger, 2023, p. 2), and Germany and Italy are protesting a ban on new combustion-engine cars (Fokuhl et al., 2023, p. 2). The EU is looking to gain an accord with the US to provide access to some benefits contained in Joe Biden's eco investment plan (Martin, 2023, p. 2), while the G7 agreed to work on German Chancellor Olaf Scholz's idea of making 'climate protection a value advantage' (Mathiesen & Von Der Burchard, 2022, p. 2) with consequent green job spin-offs and growth.

Although the EU has a quarter of global production in electric vehicles (EVs), this is under threat from Chinese hyper-production and EU tax breaks in Joe Biden's green plan, and the EU has provided little details on such skills in transition (Howarth, 2023, pp. 1–3). The Biden-led US IRA of 2022 'threatens to pull clean tech investments, and the jobs related to it, away from Europe', with Volkswagen pondering a shift to a US-based car-battery plant as it may gain a $10.54 billion subsidy (Schroder et al., 2023, p. 2). Here, the EU failed to gain agreement on reforming the carbon market, meaning the proposal will be delayed, and could benefit US jobs as companies may go outside the EU to produce goods (Spirlet, 2022, pp. 1, 4, 7). Moreover, Ford made thousands of job losses in Europe, and boss Jim Farley 'estimates that 40% fewer staff will be needed to develop the battery version' of new Ford cars (Marlow, 2023, p. 34). I now outline some green job definitions (Kozar & Sulich, 2023) and their numbers, plus the methods used in this study.

References

International Context

AccuWeather. (2022). 'Europe's brutal heat wave may be continent's worst since 1757', *AccuWeather.com,* 16th July, pp. 1–8, https://www.accuweather.com/en/weather-forecasts/europe-heat-wave-could-be-worst-in-over-200-years/1216701

Allen, G. (2017). 'Donald Trump didn't even wait for his speech to end before he announced scrapping Environmental laws', *Mirror Online,* 20th January, pp. 1–7. https://www.mirror.co.uk/news/world-news/donald-trump-didnt-even-wait-9661579

Allen, N. (2018). 'Five ways California faces disaster at the hands of Mother Nature', *The Daily Telegraph,* 25th January, p. 1.

Asthana, A., & Wintour, P. (2017). 'Trump left in cold over Paris climate agreement at end of G20 summit', *The Guardian,* 8th July, pp. 1–3. https://www.theguardian.com/world/2017/jul/08/donald-trump-paris-climate-agreement-g20-summit-us-theresa-may

Baldry, C., & Hyman, J. (2022). *Sustainable work and the climate crisis: The link between labour and climate change* (pp. 1–277). Routledge.

Baptiste, N. (2017). 'Trump Wants to Decimate Superfund. Here's Why That Is Such a Terrible Idea', *Mother Jones,* 6th April, pp. 1–4.

BBC News. (2017a). 'Hottest day since summer of 1976 in heatwave', *BBC News,* 21st June, pp. 1–5.

BBC News. (2017b). 'Government has failed to act on air pollution, says Labour', *BBC News,* pp. 1–4.

BBC News. (2017c). 'Phoenix flights cancelled because it's too hot for planes', *BBC News,* 20th June, pp. 1–7.

BBC News. (2023a). 'Climate change: July set to be world's warmest month on record', *BBC News,* Thursday 27th July, pp. 1–11, https://www.bbc.co.uk/news/science-environment-66322608

Biden, J. (2021) 'Biden-Harris Democrats - Climate: 10 Million Clean Energy Jobs', joe.biden.com, https://joebiden.com/climate-labor-fact-sheet/# pp. 1–8.

Birkett, E., Luke, A., & Bunt, P. (2022) 'Green jobs, red wall: why green industrial jobs are critical to levelling up', *Onward UK,* 16th December, pp. 1–10, https://www.ukonward.com/reports/green-jobs-red-wall/

Bloomberg News. (2020). 'The secret origins of China's 40-year plan to end carbon emissions', *Bloomberg News*, 22nd November, pp. 1–9.

Cahill, T. (2015). 'Bill Gates: Only Socialism Can Save The Climate, 'The Private Sector is Inept', *U.S. Uncut,* pp. 1–8, 26th October, http://strategicstudyindia.blogspot.com/2015/12/bill-gates-only-socialism-can-save.html

Carrington, D. (2016). 'Indoor and outdoor air pollution "claiming at least 40,000 UK lives a year"', *The Guardian,* Monday 22 February, pp. 1–3.

Carrington, D. (2017). 'UK's new air pollution plan dismissed as "weak" and 'woefully inadequate', *The Guardian,* Friday 5th May, pp. 1–4.

Channel 4. (2017). Dispatches: President Trump's Dirty Secrets, 16th January 2017.

Chung, M. (2017). 'Paris Withdrawal Doesn't Have to be the End of American Leadership on Climate', June 1, https://www.linkedin.com/pulse/paris-withdrawal-doesnt-have-end-american-leadership-climate-chung

Clark, R. (2022). 'Sunak is right to stay away from COP27', *The Spectator,* 28th October, pp. 1–3.
Coen, S. (2023). 'Biden's "environmental extremism" will decimate the car industry says Trump', *The Daily Telegraph,* Tuesday 27th June, p. 12.
Crilly, R. (2017). 'Donald Trump pulls out of Paris climate accord to "put American workers first"', *The Daily Telegraph,* 2nd June, https://www.telegraph.co.uk/news/2017/06/01/trump-pull-paris-accord-seek-better-deal/
Daily Express. (2018). 'Manchester river has most plastic pollution in the world', *Daily Express,* Tuesday 13th March, pp. 1–3.
Daily Telegraph. (2020). 'Joe Biden says "America is back" - but stresses it won't be an "Obama third term"', *The Daily Telegraph,* 25th November, p. 1.
Daily Telegraph and Agence France-Presse. (2017). 'Trump abandons Paris climate change agreement: How the business world reacted', *The Daily Telegraph,* 2nd June, https://www.telegraph.co.uk/business/2017/06/02/trump-abandons-paris-climate-change-agreement-business-world/
Extreme Temps. (2022). *Twitter/X,* Thursday 17th August, p. 1.
Farndale, E., Bonache, J., McDonnell, A., & Kwon, B. (2023). Positioning context front and Centre in international human resource management. *Human Resource Management Journal, 33,* 1–16.
Flavelle, C. (2016). 'Climate Change Is Already Forcing Americans to Move', Bloomberg UK, 31st October, pp. 1–7, https://www.bloomberg.com/opinion/articles/2016-10-31/climate-change-is-already-forcing-americans-to-move?leadSource=uverify%20wall
Gatten, E. (2023a). 'Heatwave in Europe blows cold and storms to Britain', *The Daily Telegraph,* Friday 14th July, p. 3.
Gatten, E. (2023b). 'Global warming likely to hit 1.5C tipping point by 2028', *The Daily Telegraph,* Thursday 18th May, p. 11.
Gill, V. (2023). 'Metal-mining pollution impacts 23 million people worldwide', *BBC News,* 23rd September, pp. 1–12.
Goodman, P. (2023). 'Tory net-zero sceptics sense the wind of change', *The Times,* Wednesday 26th July, p. 21.
Grover, S. (2018). 'Former coal mine could be largest solar farm in Kentucky', Treehugger.com, https://www.treehugger.com/former-coal-mine-become-largest-solar-farm-kentucky-4848913
Harrabin, R. (2023). 'Letters to the Editor: The climate threat', *The Daily Telegraph,* Monday 14th August, p. 15.
Havel, V. (1987). *Living in truth.* Faber and Faber. ISBN-10: 0571148743/ ISBN-13: 978-0571148745.

Helm, T., & Harvey, F. (2022). 'No 10 alarm as Boris Johnson plans to attend Cop27 climate summit', *The Observer,* 29th October, pp. 1–3.
Heywood, A. (2021). *Political ideologies: An introduction* (pp. 1–292). Palgrave Macmillan/Red Globe Press.
Holland, L. (2021). 'Joe Biden's climate goals already have activists breathing a sigh of relief', *Sky News,* Thursday 21st January, pp. 1–10.
Hoskins, B. (2023). 'Letters to the Editor', *The Daily Telegraph,* Tuesday 25th July, p. 15.
Houser, K. (2017). 'Germany just broke its energy record and generated 85% of electricity from renewables', *World Economic Forum/Futurism,* 18th May, pp. 1–4.
Hudema, M. (2023) *Twitter/X,* Thursday 17th August, p. 1.
Ingham, J. (2021). 'Green revolution can turbocharge UK economy and create 250,000 jobs', *Daily Express,* 8th February, pp. 1–3.
Johnson, E. M. (2015). 'Washington state sues U.S. over toxic vapours at nuclear waste site', *Reuters Environment,* 2nd September, pp. 1–2.
Kirk, A., & Scott, P. (2017). 'How high is air pollution in your city and how does it compare to the most polluted cities in the world?', *The Daily Telegraph,* 2nd August, pp. 1–4.
Kozar, L. J., & Sulich, A. (2023). Green jobs: Bibliometric review. *The International Journal of Environmental Research and Public Health, 20*, 2886. https://doi.org/10.3390/ijerph20042886
Laville, S. (2017). 'Air pollution plan cannot be delayed, high court tells government', *The Guardian,* Thursday 27th April, pp. 1–4.
Marlow, B. (2023) 'Politicians are forcing electric cars on us', *The Daily Telegraph,* Saturday 1st July, p. 34.
McGoogan, C., & Sherlock, R. (2017). 'G20 leaders roll back commitment to free trade and climate change following US pressure', *The Daily Telegraph,* pp. 1–2, 18th March, https://www.telegraph.co.uk/business/2017/03/18/g20-leaders-roll-back-commitment-free-trade-climate-change-following/
Milam, G. (2020) Thousands have died prematurely in the US due to wildfire pollution, say scientists, *Sky News,* Saturday 21st November, pp. 1–8.
Milman, O. (2016). 'Donald Trump picks climate change sceptic Scott Pruitt to lead EPA', *The Guardian,* 8th December, https://www.theguardian.com/us-news/2016/dec/07/trump-scott-pruitt-environmental-protection-agency
Monbiot, G. (2016). 'The 13 impossible crises that humanity now faces', *The Guardian,* 25th November, https://www.theguardian.com/commentisfree/2016/nov/25/13-crises-we-face-trump-soil-loss-global-collapse

Monbiot, G. (2022a). 'Fossil fuel burning once caused a mass extinction – now we're risking another', *The Guardian,* Thursday 3rd November, pp. 1–3, https://www.theguardian.com/commentisfree/2022/nov/03/fossil-fuel-burning-mass-extinction-devonian-coastline

Napolitano, J. (2017). 'The enormous risk of being silent on climate change', *Linkedin Pulse,* 22nd February, https://www.linkedin.com/pulse/enormous-risk-being-silent-climate-change-janet-napolitano/

Palmer, A. (2020). 'Amazon threatens to fire employees who speak out on climate change', *CNBC.com,* pp. 1–8.

Poschen, P. (2015). *Decent work, green jobs and the sustainable economy* (pp. 1–182). Greenleaf Publishing Limited/International Labour Organization (ILO).

Reuters News Agency. (2017). 'The US will join climate talks despite quitting Paris accord', *The Daily Telegraph,* 4th August, pp. 1–4.

Sabur, R. (2023). 'Defiant Trump says he will run for president even if sentenced in a criminal case', *The Daily Telegraph,* 29th July, p. 14, https://www.telegraph.co.uk/world-news/2023/07/28/donald-trump-run-president-even-if-sentenced-criminal-case/

Shah, J. (2017). 'Executive Order to Create Energy Independence be P****** in the Wind', *LinkedIn,* 28th March, pp. 1–4.

Shipman, A. (2020). 'Obama warned not to "underestimate Biden's ability to f*** things up"', *Daily Express,* 1st December, pp. 1–3.

Simpson, A. (2016). 'Businesses no longer have an excuse not to disclose their climate risks', *The Guardian,* 29th December, https://www.theguardian.com/sustainable-business/2016/dec/29/business-climate-risk-disclosure-bloomberg

Singh, D. (2018). 'Green Jobs Horizon Expends in Rapid Climate Changes Era', *ABC Live India,* 1st June, pp. 1–3.

Sky News. (2020). 'Paris climate deal was "designed to kill the American economy", Donald Trump claims', *Sky News,* Monday 23rd November, pp. 1–7, https://news.sky.com/story/paris-climate-deal-was-designed-to-kill-the-american-economy-donald-trump-claims-12139272

Sky News. (2023). 'Deaths linked to extreme heat on hottest days soar', *Sky News,* Friday 22nd September, p. 1.

Sky News Live. (2023). Sophie Ridge on Sunday, *Sky News Live,* Sunday 2nd July.

Smith, H. (2023). 'People feel unprotected': Greeks lose faith in state after Storm Daniel and summer of wildfires', *The Guardian,* Saturday 16th September, pp. 1–6.

Sopel, J. (2021). 'Biden 100 days: what we all got wrong about him', *BBC News,* 27th April, pp. 1–12. https://www.bbc.co.uk/news/world-us-canada-56903805

Spencer, M. (2017) 'How American policy on the Environment is losing the goodwill of the Planet', *LinkedIn,* pp. 1–5.

Squires, N., Badcock, J., & Song, V. (2023) 'Killer heatwave descends on Europe as temperatures reach 40C', *The Daily Telegraph,* Thursday 13th July, p. 13.

The Economist. (2023a). 'America's energy transition: The irresistible nation', *The Economist,* April 8th–14th, pp. 31–33.

The Economist. (2023b). 'Leader: Big, green and mean', *The Economist,* 4th February, p. 9.

The Economist. (2023c). 'United States: Green v green – America needs a new environmentalism', *The Economist,* 4th February, pp. 33–35.

Trump, D. (2016). *Twitter/X,* 22nd November, p. 1.

United Nations News (UN News). (2020). 'UN builds momentum for restoring forests as the world enters key decade for ecosystems', *United Nations News,* 28th October, pp. 1–3.

Valero, A, Li, J., Muller, S, Riom, C., Nguyen-Tien, V., & Draca, M. (2021) 'Policy brief: Are "green" jobs good jobs? Lessons from the UK and EU to inform labour market transitions', Grantham Research Institute on Climate Change and the Environment/Centre for Climate Change Economics and Policy, *London School of Economics and Political Science,* October, pp. 1–8, https://www.lse.ac.uk/granthaminstitute/publication/are-green-jobs-good-jobs-how-lessons-from-the-experience-to-date-can-inform-labour-market-transitions-of-the-future/

Warner, J. (2023). 'Heat pumps could be about to topple their first government', *The Daily Telegraph,* Wednesday 31st May, p. 21.

EU Green Policies

BBC News. (2023b). 'France bans short-haul flights to cut carbon emissions', *BBC News,* 23rd May, pp. 1–12.

Beirao, F. (2022). *Twitter/X,* Monday 21st February, p. 1.

Business Green. (2022). *Twitter/X,* Thursday 23rd June, p. 1.

Capion, K. (2022). *Twitter/X,* Saturday 25th June, p. 1.

Carbon Brief. (2022). *Twitter/X,* Monday 6th March, p. 1.

Carbon Tracker. (2023). *Twitter/X,* Wednesday 19th July, p. 1.

Carrington, D. (2022). *Twitter/X,* Friday 8th July, p. 1.

Connolly, T. (2022). *Twitter/X,* Wednesday 18th May, p. 1.
Covering Climate. (2023). *Twitter/X,* Saturday 22nd July, p. 1.
ECIU. (2022). *Twitter/X,* Saturday 18th March, p. 1.
ECIU. (2023). *Twitter/X,* Friday 7th April, p. 1.
Electrify Heat. (2022). *Twitter/X,* Thursday 23rd June, p. 1.
Ember Climate. (2022a). *Twitter/X,* Monday 28th February, p. 1.
Ember Climate. (2022d). *Twitter/X,* Monday 30th May, p. 1.
Ember Climate. (2022j). *Twitter/X,* Tuesday 7th June, p. 1.
Ember Climate. (2022l). *Twitter/X,* Saturday 25th June, p. 1.
Europe Beyond Coal. (2021). *Twitter/X,* Tuesday 14th December, p. 1.
Europe Beyond Coal. (2022). *Twitter/X,* Thursday 12th May, p. 1.
European Commission. (2021a). *Twitter/X,* Saturday 27th November, p. 1.
European Commission. (2021b). *Twitter/X,* Wednesday 15th December, p. 1.
European Commission. (2022a). *Twitter/X,* Tuesday 4th January, p. 1.
European Commission. (2022b). *Twitter/X,* Sunday 16th January, p. 1.
European Commission. (2022c). *Twitter/X,* Sunday 30th January, p. 1.
European Commission. (2023a). *Twitter/X,* Wednesday 29th March, p. 1.
European Commission. (2023b). *Twitter/X,* Sunday 30th April, p. 1.
European Commission. (2023c). 'Statement by President von der Leyen on the Green Deal Industrial Plan', *European Commission,* 1st February, pp. 1–8, https://ec.europa.eu/commission/presscorner/detail/en/statement_23_521
European Commission. (2023d). *LinkedIn,* Thursday 10th August, p. 1.
Evans, S. (2022a). *Twitter/X,* Tuesday 8th March, p. 1.
Hahn, J. (2023). *Twitter/X,* Tuesday 28th March, p. 1.
Happaerts, S. (2021). *Twitter/X,* Tuesday 14th December, p. 1.
Happaerts, S. (2022). *Twitter/X,* Monday 8th August, p. 1.
Harvey, F. (2023). *Twitter/X,* Tuesday 3rd January, p. 1.
Hudema, M. (2022). *Twitter/X,* Friday 14th October, p. 1.
Johnson and Johnson. (2022). *Twitter/X,* Tuesday 8th February, p. 1.
Mathis, W. (2022). *Twitter/X,* Wednesday 10th August, p. 1.
Murray, J. (2022). 'Historic decision': MEPs vote to end sale of petrol and diesel cars, but challenges remain for EU climate plans', *Business Green,* 9th June, pp. 1–7, https://www.businessgreen.com/news-analysis/4050998/historic-decision-meps-vote-end-sale-petrol-diesel-cars-challenges-remain-eu-climate-plans
Polman, P. (2022). *Twitter/X,* Saturday 5th February, p. 1.
Ralston, J. (2022). *Twitter/X,* Wednesday 22nd June, p. 1.
Rathi, A. (2021). *Twitter/X,* Monday 22nd November, p. 1.

Razzouk, A. (2021). *Twitter/X,* Saturday 20th November, p. 1.
Razzouk, A. (2022). *Twitter/X,* Tuesday 30th August, p. 1.
Razzouk, A. (2023). *Twitter/X,* Sunday 30th April, p. 1.
Spurway, P. (2022). *Twitter/X,* Thursday 27th January, p. 1.
Taylor, K. (2022a). *Twitter/X,* Wednesday 11th May, p. 1.
Timmermans. (2022). *Twitter/X,* Thursday 23rd June, p. 1.
Ukraine World. (2022). *Twitter/X,* Wednesday 26th October, p. 1.
Van der Leun, K. (2022a). *Twitter/X,* Tuesday 1st February, p. 1.
World Economic Forum. (2022). *Twitter/X,* Tuesday 13th September, p. 1.

EU Criticism

Ainger, J. (2021). 'Coal Plants in Southeast Europe Killing Thousands, Report Says', *Bloomberg News,* 6th September, pp. 1–8, https://www.bloomberg.com/news/articles/2021-09-06/coal-plants-in-southeast-europe-killing-thousands-report-says
Axelsson, I. (2022). *Twitter/X,* Friday 21st January, p. 1.
Blake, R. (2021). *Twitter/X,* Thursday 11th November, p. 1.
Bloomberg Green. (2023). *Twitter/X,* Tuesday 4th July, p. 1.
Boffey, D. (2022) 'EU faces legal challenge over plan to fast-track gas projects', *The Guardian,* Tuesday 7th June, pp. 1–3. https://www.theguardian.com/environment/2022/jun/07/eu-faces-legal-challenge-over-plan-to-fast-track-gas-projects
Brightwell, R. (2022). *Twitter/X,* Tuesday 8th February, p. 1.
CAN Europe. (2021). *Twitter/X,* Wednesday 17th November, p. 1.
Carbon Tracker. (2022a). *Twitter/X,* Tuesday 15th February, p. 1.
Carbon Tracker. (2022b). *Twitter/X,* Sunday 20th March, p. 1.
Centre for Research on Energy and Clean Air. (2022). *Twitter/X,* Friday 14th October, p. 1.
Chatham House. (2022). *Twitter/X,* Saturday 2nd July, p. 1.
Clean Air Task Force. (2023). *Twitter/X,* Friday 12th May, p. 1.
Cremona, L. (2022). *Twitter/X,* Tuesday 25th January, p. 1.
Ember Climate. (2021a). *Twitter/X,* Thursday 25th November, p. 1.
Ember Climate. (2021b). *Twitter/X,* Friday 3rd December, p. 1.
Ember Climate. (2021c). *Twitter/X,* Wednesday 8th December, p. 1.
Ember Climate. (2022b). *Twitter/X,* Tuesday 1st February, p. 1.
Ember Climate. (2022c). *Twitter/X,* Wednesday 2nd February, p. 1.
Ember Climate. (2022e). *Twitter/X,* Tuesday 25th January, p. 1.

Ember Climate. (2022h). *Twitter/X,* Sunday 1st May, p. 1.
Ember Climate. (2023). *Twitter/X,* Tuesday 27th June, p. 1.
European Commission. (2021). *Twitter/X,* Sunday 21st November, p. 1.
Evans, S. (2022b). *Twitter/X,* Friday 20th May, p. 1.
Genicot, D. (2023). *Twitter/X,* Friday 3rd March, p. 1.
Green Finance Institute. (2021). *Twitter/X,* Tuesday 23rd November, p. 1.
Green Party. (2022). *Twitter/X,* Tuesday 15th February, p. 1.
Harrabin, R. (2022). *Twitter/X,* Friday 6th May, p. 1.
Hoepner, A. (2022). *Twitter/X,* Wednesday 5th January, p. 1.
Industriall Europe. (2023) *Twitter/X,* Thursday 9th March, p. 1.
International Resource Panel. (2022). *Twitter/X,* Monday 18th April, p. 1.
Jones, D. (2022). *Twitter/X,* Friday 14th January, p. 1.
McVeigh, K. (2022). 'EU slammed over failure to protect marine life from "destructive" fishing', *The Guardian,* Monday 12th September, pp. 1–5, https://www.theguardian.com/environment/2022/sep/12/eu-slammed-over-failure-to-protect-marine-life-from-destructive-fishing#:~:text=EU%20slammed%20over%20failure%20to%20protect%20marine%20life%20from%20'destructive'%20fishing,-This%20article%20is&text=The%20waters%20of%20the%20EU,than%20200%20scientists%20and%20conservationists
Morgan, S. (2022). *Twitter/X,* Tuesday 1st February, p. 1.
O'Leary, N. (2021). 'Greta Thunberg takes aim at EU's Common Agricultural Policy', *The Irish Times,* Wednesday 31st March, pp. 1–11, https://www.irishtimes.com/news/world/europe/greta-thunberg-takes-aim-at-eu-s-common-agricultural-policy-1.4525494
Politico Europe. (2022). *Twitter/X,* Tuesday 4th January, p. 1.
Rall, K. (2021). *Twitter/X,* Thursday 25th November, p. 1.
State of Global Air. (2022). *Twitter/X,* Wednesday 4th May, p. 1.
Taylor, K. (2022b). *Twitter/X,* Thursday 13th October, p. 1.
The Economist. (2021). 'Nuclear energy united Europe. Now it is dividing the club', *The Economist,* 30th October, pp. 1–5, https://www.economist.com/europe/2021/10/30/nuclear-energy-united-europe-now-it-is-dividing-the-club?utm_medium=cpc.adword.pd&utm_source=google&ppccampaignID=18156330227&ppcadID=&utm_campaign=a.22brand_pmax&utm_content=conversion.direct-re-sponse.anonymous&gclid=Cj0KCQjw0IGnBhDUARIsAMwFDLkR3QrcDU6MrG6xPQtq2_6dvnt15J2Xy8PaN8xuMRW3dgpa3HzU3YUaAuUREALw_wcB&gclsrc=aw.ds
Thunberg, G. (2022a). *Twitter/X,* Thursday 3rd February, p. 1.

Thunberg, G. (2022b). *Twitter/X,* Monday 20th June, p. 1.
Thunberg, G. (2022c) *Twitter/X,* Wednesday 6th July, p. 1.
Van der Leun, K. (2022b). *Twitter/X,* 1st February, p. 1.
Wallace-Wells. (2022). *Twitter/X,* Monday 4th July, p. 1.

EU Green Jobs

Capps, K. (2022). 'Why Lithuania Leads Europe in Mobilizing Green Building Retrofits', *Bloomberg News,* 21st September, pp. 1–12, https://www.bloomberg.com/news/articles/2022-09-21/why-lithuania-leads-europe-in-mobilizing-green-building-retrofits

Carbon Brief. (2023). 'Q&A: How the EU wants to race to net zero with 'Green Deal Industrial Plan', *Carbon Brief,* 17th March, pp. 1–21, https://www.carbonbrief.org/qa-how-the-eu-wants-to-race-to-net-zero-with-green-deal-industrial-plan/#:~:text=A%20key%20goal%20of%20the,potential%20best%20practices%E2%80%9D%20for%20finance

European Parliament. (2022). 'Parliament backs boost for renewables use and energy savings', *European Parliament News,* 14th September, pp. 1–4, https://www.europarl.europa.eu/news/en/press-room/20220909IPR40134/parliament-backs-boost-for-renewables-use-and-energy-sav-ings#:~:text=On%20Wednesday%2C%20MEPs%20voted%20to,under%20its%20%22RepowerEU%22%20package

Fisher, J. (2022). 'EU reveals its plans to stop using Russian gas', *BBC News,* 18th May, pp. 1–10, https://www.bbc.co.uk/news/science-environment-61497315

Fokuhl, J., Albanese, C., & Ainger, J. (2023). 'Germany, Italy Protest Strict EU Plan for Combustion-Engine Ban', *Bloomberg Law News,* 1st March, pp. 1–8, https://news.bloomberglaw.com/environment-and-energy/italy-and-germany-threaten-eu-plan-to-ban-combustion-engines

Howarth, T. (2023). 'Here's a glimpse of the EU response to the Inflation Reduction Act', *Green Biz,* 6th February, pp. 1–4, https://www.greenbiz.com/article/heres-glimpse-eu-response-inflation-reduction-act

IRENA. (2023). *Twitter/X,* Sunday 30th April, p. 1.

Martin, E. (2023). 'EU Seeks Trade Deal With US to Unlock Electric Car Benefits', *Bloomberg News,* 3rd March, pp. 1–7, https://www.bloomberg.com/news/articles/2023-03-03/eu-seeks-preliminary-fta-like-deal-with-us-as-soon-as-next-week?in_source=embedded-checkout-banner https://www.bloomberg.com/news/articles/2023-03-03/eu-seeks-preliminary-fta-like-deal-with-us-as-soon-as-next-week?in_source=embedded-checkout-banner

Mathiesen, K., & Von Der Burchard, H. (2022). 'G7 buys into Scholz's Club idea', *Politico,* 28th June, pp. 1–10, https://www.politico.eu/article/g7-buys-into-scholzs-climate-club-idea/#:~:text=The%20club%20would%20help%20countries,advantage%20and%20not%20a%20disadvantage
Mathis, W. (2023). 'Solar Beats Coal in Europe for First Time – But There's a Glitch', *Bloomberg News,* 3rd June, pp. 1–7, https://www.bloomberg.com/news/articles/2023-06-03/solar-beats-coal-in-europe-for-first-time-but-there-s-a-glitch
Monbiot, G. (2022b). 'Tory MPs call the green transition "unaffordable". Europe is proving that is a lie', *The Guardian,* Wednesday 23rd March, pp. 1–6, https://www.theguardian.com/commentisfree/2022/mar/23/european-countries-climate-crisis-deniers-green-schemes
Reierson, A., & Ainger, J. (2023). 'Wind Investment in Europe Is Plummeting When It's Most Needed', *Bloomberg News,* 30th March, pp. 1–7, https://www.bloomberg.com/news/articles/2023-03-30/wind-investment-in-europe-is-plummeting-when-it-s-most-needed
Rolt, A. (2022). 'IEA: Heat pump sales surge 40 per cent in Europe to three million units in 2022', *Business Green,* 3rd April, pp. 1–8, https://www.businessgreen.com/news/4111417/iea-heat-pump-sales-surge-cent-europe-million-units-2022
Schroder, P., Bergsen, P., & Barrie, J. (2023). 'Europe's pursuit of securing critical raw materials for the green transition', *Chatham House,* 4th April, pp. 1–6, https://www.chathamhouse.org/2023/04/europes-pursuit-securing-critical-raw-materials-green-transition-0
Spirlet, T. (2022). '"Raucous day" Von der Leyen's energy reform "defeated" in split EU vote – "Victory" for US', *Daily Express,* Thursday 9th June, pp. 1–8, https://www.express.co.uk/news/world/1623222/EU-news-ursula-von-der-leyen-energy-reform-climate-change-usa-vn

2

Green Jobs, Theoretical Framing and Methodological Approach

Abstract In this chapter, I critically discuss development in UK green jobs in general and among the Anglosphere and then the study focus, theoretical framing and methodology. Here, I state my spotlight and approach, data sources and purposes, discursive approach and review procedure.

Keywords UK • Green • Jobs • Theory • Framing • Methods • Data • Sources • Review • Procedure

Green Jobs

A question that arises when thinking about green jobs is how national, regional and global patterns of green employment appear, and some relevant research on them has been done by the International Labour Organization (ILO) (see Poschen, 2015). However, frustration occurs on gaining such useful data in the US, for example, as figures held there seemingly only relate to the state level,[1] meaning that a US-wide picture

[1] According to the Baker Library leader at Harvard Business School.

D. W. S. Renwick, *Work, Politics and the Green Industrial Revolution*,
https://doi.org/10.1007/978-3-031-53067-8_2

of green job patterns is not easy to build. Indeed, some authors also note that 'there is not yet an agreed definition of what a green job is, making comparing existing research findings difficult' too (Valero et al., 2021, p. 2). Yet there has also been some useful work done in the US via O*NET, their occupational classification database, which examines the 'green transition' and identifies three categories of occupations likely to change or emerge in green economy sectors, namely: 'Green Increased Demand', 'Green Enhanced Skills' and 'Green New and Emerging'. All three such O*NET categories 'offer a different way to look at the moving landscape of job and skills needs among various occupations as the transition continues' (Green Jobs Taskforce, 2021, pp. 15–16).

Nonetheless, for governments and organizational stakeholders, progress in more environmentally friendly employment saw prior US reports from the Obama administration pledging to create five million US-based green jobs (Shudworth, 2017b). Such US green job growth may now be driven up by the cheaper cost of solar energy, as some US companies view solar as an unlimited energy source and a 'no-brainer investment' (Shudworth, 2017a). In North Carolina, Amazon is building a wind farm, the Georgia metro system has seen huge growth, US solar employment in 2016 contained 260,077 workers nationwide, and a broad labour impact including 789,000 direct and indirect US jobs has occurred too (Shah, 2017, p. 3). Further, more solar energy jobs were reported 'than in oil, natural gas and coal in the US' (Simpson, 2016, p. 2), never mind now in the pro-green Biden era (see below).

However, the future for some US coal workers also seems problematic, as coal-fired profitability is declining, US coal use has dropped sharply since 2007, the world's largest private sector coal company Peabody Energy went bankrupt in 2016 and only 150,000 workers had jobs in the domestic US coal industry in 2015 (Pearce, 2016, p. 2). Yet the US solar industry can also absorb some coal-industry redundancies and offer full-time careers, as coal operation engineers could retrain as manufacturing technicians in solar, while explosives employees, ordinance handlers and blasters in the coal industry could take their safety knowledge and skills to gain extra training and become solar technicians, and janitors in coal might become low-skilled mechanical assemblers in solar too (Pearce, 2016, p. 3). Additionally, structural engineers in the coal industry would not need much new schooling to work in solar, and some coal worker retraining may only need short courses or on-the-job training, meaning a relatively minor investment of $180 million to $1.8 billion (based on best and worst-case scenarios) in retraining would allow many US coal workers to switch to solar-related positions (Pearce, 2016, pp. 3–4). However, for US coal workers, upskilling costs facilitating their transfer into the emerging US solar industry are high too (Pearce, 2016, pp. 1–3).

Such pre-Joe Biden developments have now been strengthened in the US, as the Biden administration has seemingly mandated that green issues are considered in all US government agencies and departments 'through a climate change lens', and publicized huge plans to generate ten million US-based green energy jobs (Biden, 2021). Such figures build on prior estimates that the US green economy was worth some $1.3 trillion and employed more than 4% of working-age Americans, i.e. ten times more than in fossil fuels (Baldry & Hyman, 2022, p. 87). But one fly in the ointment of new US green job plans lies in nuclear power, which while huge as it has over 90 nuclear reactors spread across 28 states, needs new policy movement, as if not it will retire more than half the US nuclear fleet before 2030 (Hamer, 2021, p. 24). Doing so means US job losses, redundancies, retraining, reskilling and upskilling in nuclear power seem needed.

I now briefly detail developments in green jobs among some key Anglo countries, not just to detail numbers of such green jobs arising (as scholars can add more detail from relevant global databases and other sources such as Shah, 2017, p. 3), but to chart the wider international context within which UK government green job creation occurs.

Green Job Numbers in the Anglosphere

In terms of overall numbers, the renewable energy sector employs over eight million workers globally, and big green job gains have been recorded in it (Singh, 2018, p. 2). Here, in the US 2.7 million jobs have been generated in the environmental goods and services sector (Singh, 2018, p. 2), and the EU had 4.2 million people employed in the green industry (Demogenes, 2020), while in the UK, which is the country context and subject of this book, the Green Jobs Taskforce (GJT) Executive Summary Introduction says that the British government has 'set an ambition for two million green jobs in the UK by 2030' and in July 2021 reported that Britain already had 400,000 UK-based green jobs (Green Jobs Taskforce, 2021, pp. 5–6). However, other alternative and more critical sources conversely reported no net gains in UK green jobs overall (Harrabin, 2021). Further, analysis by the Energy and Climate Intelligence Unit (ECIU) drawing on CBI Economics and The Data City figures reports 840,000 (3.2%) UK jobs being supported by businesses in the net zero economy (Energy and Intelligence Climate Unit (ECIU), 2023, p. 2). Therefore, figures on UK green job creation are (cf. Edwards, 1979) 'contested terrain'. I now outline the focus of this study, the theoretical lens adopted and the methods used.

Focus, Theoretical Framing and Methodology

Drawing on Oswick (2012), I adopt discourse analysis herein to study how meanings are generated and dominate through language-in-use and to assess the meaning of words concerning their context (cf. Keenoy et al., 1997). I explore the socio-political context of the UK Green Job Taskforce (GJT) using related texts published by official British government sources, recent UK conservative politicians (Fairclough, 1995; Lee, 2021, p. 9), British-based trade union leaders and both orthodox and unorthodox UK media sources (cf. Bell et al., 2016; Oswick, 2012, pp. 475–476). In doing so, I examine different analytical levels, ranging from a micro-level deconstruction of language use via a reading of UK government and leader texts to a macro-focus meta-discourse of conservative political ideas and ideology from Goodwin (2014) to 'uncover patterns in discursive themes' (Oswick, 2012, pp. 476–477, 481–482). In using a critical epistemology and deconstruction method, I offer a close reading and analysis of the GJT while acknowledging that it represents the interests of the British state and that I tend to focus on elite voices in, around and influencing the UK government regarding it (cf. Oswick, 2012, p. 483). As such, I seek to reveal patterns of oversimplification, contradictions, partiality and hierarchical ordering in the GJT (cf. Oswick, 2012, p. 485), to give current UK political discourse on green employment attention and bring issues regarding it 'to life' (cf. Howard-Grenville, 2021, p. 9), by using the GJT as an explanatory case in new British government green jobs policy (cf. Sandberg and Alvesson, 2021; Yin, 1994).

Spotlight and Approach

This research monograph is an initial, scoping and exploratory investigation which for academics extends the now popular topic and conversations surrounding green HRM (Memon et al., 2022) forward by critically analysing the UK Green Jobs Taskforce (GJT) report through the less familiar theoretical lens of conservative political ideas, i.e. on and around its own political terms, as a piece of key 'concrete' UK government policy (cf. Dobbins, 2022, p. 34). For policy experts and practitioners, it is intended as an academic resource book to indicate the current state of play on the topic of green jobs in the UK. In doing so, I act to understand, and bring in, how green regulatory contexts established, detailed and enacted by national governments help or hinder businesses in green job creation, i.e. how politics shapes and impacts HRM in general and green HRM in particular. Thus, this research book is about adding to knowledge on the political context in HRM and Management studies, which is becoming more popular, as others now research such activities (see Gounopoulos et al., 2022; Schuessler et al., 2023: Sun et al., 2023, pp. 23–24), and related political, gender and social class issues related to climate change and sustainability too (e.g. Guerci et al., 2023; Maher et al., 2022; Mittiga, 2021). Here, me specifically using a theoretical lens derived from political studies is a novel development in HRM and Management studies, as doing so focuses on issues of government regulation and context external to, and outside of organizations, brings them into organizations, and therefore makes such issues front-and-centre within both HRM and Management studies too (cf. Cooke, 2017; D'Cruz et al., 2021; Farndale et al., 2023; Rodriguez et al., 2017; Vincent et al., 2020).

Due to the very recent, ongoing and 'live' nature of this topic, I use several somewhat unorthodox sources to draw it together, such as British media and press reports and social media including Twitter/X before Elon Musk's takeover of it (Smith, 2023; Stokel-Walker, 2023), alongside more orthodox sources such as official UK government documents, opinion pieces and academic works. Here, I do not claim any definitive knowledge on all things happening in the UK green jobs domain or sphere,[2] as researching it is a brand-new topic for me. Instead, interested readers could consult the works of top journalists such as Jo Alsop, Damian Carrington, Roger Harrabin, George Monbiot and James Murray, plus leading academics such as Professor Vera Trappmann and green activists like Caroline Lucas MP and Jonathon Porritt among others, for more relevant, factual and historical data. As such, this research monograph is not an in-depth investigation of the 'just transition' and issues of 'social justice' for UK workers moving from fossil fuel jobs to green(er) renewable energy-related ones, as, while I recognize these are crucially important issues for all relevant stakeholders involved, this research either has already been done by other academics or is an ongoing research undertaken in a comparative frame elsewhere. For example, interested scholars may wish to read the many projects already done in these two social justice areas by Professor Vera Trappmann at the University of Leeds, the UK, at: https://business.leeds.ac.uk/departments-work-employment-relations/staff/544/professor-vera-trappmann. Instead, the research volume herein only concerns the general political aspects, concerns, ideas, issues and consequences arising surrounding UK green job generation and eco-employment in Britain.

Data Sources and Purposes

Inspired by Lee (2021), the UK-based documents I interrogated such as policy statements and strategic plans acted as organizational prescriptions of what they 'would like to happen in the future'. Alongside my use of

[2] Instead, readers could consult these official UK government policies: the *Prime Minister's Ten Point Plan for a Green Industrial Revolution*, the *Energy White Paper, North Sea Transition Deal, Industrial Decarbonisation Strategy, Transport Decarbonisation Plan, Hydrogen Strategy, Commitment to Decarbonise the UK's Electricity System by 2035, and the Net Zero Research and Innovation Framework* (see GOV.UK, 2021a, p. 6; 2021b).

formal UK government notes, notices and external documentary sources like trade magazines, business press commentaries and pressure group publications, other international, national and local government documents were materials that provided key details of how the UK government 'is managing its operations' and how other actors view their management too (cf. Lee, 2021, pp. 1–2). Such documents provided very useful secondary evidence, as their purpose was other than this study and predated it, and thus provide an important limitation on what I can report, as all I can portray 'is what the document – or the author of the document – is representing' (Lee, 2021, p. 3). Ergo, I recognize and accept that the meaning they intended to convey as authors may well differ from my view (and those of others) and 'could be influenced by the regulatory and institutional setting including language, belief systems and legal requirements' (Lee, 2021, p. 4). Indeed, my analysis herein could help advance my understanding of conservative political ideas and theory, as my interpretation of the author's intentions behind an emerging series of events (Lee, 2021, p. 6). In doing so, I seek to help explain why the GJT is as it is (or appears to be) because my interest in such labour history includes researching documentary sources (Lee, 2021, pp. 7–8). In taking this historical view, I use a sequencing approach to gain evidence of contemporary and contextual political events and show purpose via a 'historical trawl' with me as a commentator having a narrative voice (Lee, 2021, p. 9).

As per Oswick (2012), I use a critical discourse analysis (CDA) to focus on the socio-political context of the discursive event of the GJT, to understand how meanings are generated, and which ones hold sway (Oswick, 2010, 2012). I focus on contextual aspects (Keenoy et al., 1997), to comprehend and showcase the GJT so I can query powerful interests and raise concerns relating to UK political governance (cf. Oswick, 2012, pp. 474, 477). Here, I accept that delineating levels does not involve absolutes, because I 'combine different discursive approaches' emphasizing different levels 'to undertake forms of cross-level analysis' (Oswick, 2012, pp. 476–477). My 'mode of discursive engagement' is CDA, where I aim to interrogate 'how and why certain views and accounts dominate while others are suppressed or silenced' (Oswick, 2012, p. 477).

Discursive Approach

My method(s) to carry out a CDA include narrative analysis, as the data I used were gathered in already-known written texts, and not in interviews, observations or autoethnographies (Oswick, 2012, p. 478). I use deconstruction after Jacques Derrida, as a 'micro-level discursive technique' involving a close reading of the GJT, to surface complexity 'as a critical mode of engagement', with a focus on: 'examples used (in terms of partiality); issues of absence or presence within the text (what is included and what has excluded); aspects of hierarchical ordering; and discrepancies within the text' (Oswick, 2012, p. 479). My CDA draws on the works of Norman Fairclough (1992, 1995, 2003), in using a critical epistemology to make the power relationships of UK conservative politicians and conservative political ideas and theory 'visible through analysis' (Fairclough, 1995, p. 97), where my focus is on power issues and ideology and how texts 'are influenced by wider contextual factors and vice versa' (Oswick, 2012, p. 481). In doing so, I recognize that my choosing the GJT text itself is wholly partial and 'represents dominant interests', such as UK government policy, and my deconstruction primarily focuses on 'privileged voices', such as UK government, University academics, policy experts and journalists. As such, I accept that I may be depreciating some other viewpoints as secondary ones (Oswick, 2012, p. 483), e.g. those by Greenpeace, the UK Green Party, the UK Labour Party and other media critics.

Review Procedure

In compiling the information and data I use herein, I followed the steps outlined in the Preferred Reporting Items for Systematic Reviews and Meta-Analyses (PRISMA) 2020 checklist, which helps authors clearly and transparently report on what they do and find (Page et al., 2021). Specifically, I identified relevant eligibility criteria, which included all UK press, media, social media, reports and articles published on the GJT, British politics and green jobs from around late 2020 (when Boris

Johnson's original Ten Point Green Jobs Plan was released) up until the time of writing of mid to late 2023. I then identified information sources via searching relevant online academic databases to ensure comprehensiveness, and performed searches using keywords such as: 'UK', 'Great Britain', 'Britain', 'England', 'Scotland', 'Wales', 'Northern Ireland', 'green', 'eco', 'ecological', 'environmental', 'jobs', 'employment', 'work', 'task-force', 'decent', 'politics', 'political', 'power', 'conflict', 'state', 'society', 'government', 'governance', 'Parliament', 'conservative', 'MP', 'Tory', 'Tories', 'labour', 'liberal', 'democrats', 'green party', 'green groups', 'NGO' and 'think tank'. This process yielded 694 relevant articles, media and news items and reports which I sifted using the three UK Research Excellence Framework (REF) criteria of originality, rigour and relevance, to form the database I used. I write this text in the relatively new style and tradition of academic political narrative analysis recently popularized in Britain by the UK-based expert Professor Mark Galeotti (see Galeotti, 2019, 2020, 2022a, b among others). I now provide some background context to the British GJT.

References

Baldry, C., & Hyman, J. (2022). *Sustainable work and the climate crisis: The link between labour and climate change* (pp. 1–277). Routledge.

Bell, E., Kothiyal, N., & Willmott, H. (2016). Methodology-as-technique and the meaning of rigor in globalized management research. *British Journal of Management, 28*(3), 534–550.

Biden, J. (2021). 'Biden-Harris Democrats - Climate: 10 Million Clean Energy Jobs', pp. 1–8, joe.biden.com: https://joebiden.com/climate-labor-fact-sheet/#

Cooke, F. L. (2017). Provocation paper: Concepts, contexts, and mindsets: Putting human resource management research in perspectives. *Human Resource Management Journal, 28*(1), 1–13. https://doi.org/10.1111/1748-8583.12163

D'Cruz, P., Delannon, N., McCarthy, L., Kourula, A., Moon, J., Spence, L. J., & Gond, J. P. (2021). Human relations special issue call for papers: Contesting social responsibilities of business: Experiences in context. *Human Relations*, 1–7. https://journals.sagepub.com/page/hum/call-for-papers

Demogenes, B. (2020). 'EU Focus on the Bioeconomy Holds Great Promise for the Growth of Green Jobs', *Climate Scorecard,* 26th September 2020, https://www.climatescorecard.org/2020/09/eu-focus-on-the-bioeconomy-holds-great-promise-for-the-growth-of-green-jobs/

Dobbins, T. (2022). 'Good work: policy and research on the quality of work in the UK', House of Commons Library Research Briefing, 6th June, pp. 1–42, at: https://commonslibrary.parliament.uk/research-briefings/cbp-9561/

Edwards, R. (1979). *Contested terrain – The transformation of the workplace in the twentieth century*. Basic.

Energy & Intelligence Climate Unit (ECIU). (2023). 'Europe Steels the Deal', *ECIU,* 7th March, pp. 1–10, https://eciu.net/analysis/reports/2023/europe-steels-the-deal

Fairclough, N. (1992). *Discourse and social change*. Polity Press.

Fairclough, N. (1995). *Critical discourse analysis: The critical study of language* (pp. 1–263). Longman. ISBN 0-582-219809.

Fairclough, N. (2003). *Analysing discourse: Textual analysis for social research*. Routledge.

Farndale, E., Bonache, J., McDonnell, A., & Kwon, B. (2023). Positioning context front and center in international human resource management. *Human Resource Management Journal, 33*, 1–16.

Galeotti, M. (2019). *The Vory: Russia's super mafia* (pp. 1–326). Yale University Press. ISBN 978-0-300-24320-8.

Galeotti, M. (2020). *A short history of Russia: From the pagans to Putin* (pp. 1–190). Ebury Press/Penguin Random House. ISBN 978-1-52910-638-1.

Galeotti, M. (2022a). *Putin's wars: From Chechnya to Ukraine* (pp. 1–384). Osprey/Bloomsbury Publishing. ISBN 978-1-4728-4754-6.

Galeotti, M. (2022b). *The weaponisation of everything: A field guide to the new way of war* (pp. 1–235). Yale University Press. ISBN 978-0-3000-25344-3.

Goodwin, B. (2014). *Using political ideas (sixth edition)* (pp. 1–480). John Wiley & Sons Ltd.

Gounopoulos, D., Loukopoulos, G., Loukopoulos. P., & Wood, G. (2022). 'Corporate Political Activities and the SEC's Oversight Role in the IPO Process', *Journal of Management Studies,* (Early View), 24 November 2022, https://onlinelibrary.wiley.com/doi/full/10.1111/joms.12892

Green Jobs Taskforce. (2021). 'Green Jobs Taskforce: report to Government, Industry and the Skills Sector', *Department for Business, Energy & Industrial Strategy GOV.UK,* pp. 1–82, published and accessed 14th July 2021, https://www.gov.uk/government/publications/green-jobs-taskforce-report

Guerci, M., Hauff, S., Panichella, N., & Radaelli, G. (2023). Sustainable HRM and class-based inequality. *Personnel Review, 52*(5), 1597–1613. https://www.emerald.com/insight/content/doi/10.1108/PR-10-2021-0772/full/html

Hamer, A. (2021). 'Nuclear's new deal: Is nuclear power necessary for a net zero future', *Investors Chronicle,* 4-10th June, https://www.investorschronicle.co.uk/news/2021/06/03/nuclear-s-new-deal/

Harrabin, R. (2021). 'Keir Starmer attacks government record on green jobs'. *BBC News,* 1–7, published 8th August 2021, accessed 12th August 2021, https://www.bbc.co.uk/news/uk-politics-58088236

Howard-Grenville, J. (2021). Caring, courage, and curiosity: Reflections on our roles as scholars in organizing for a sustainable future. *Organization Theory, 2,* 1–16.

Keenoy, T., Oswick, C., & Grant, D. (1997). Organizational discourses: Text and context. *Organization, 4*(2), 147–157.

Lee, B. (2021). *How do I collect documentary evidence?* (pp. 1–128). Edward Elgar. ISBN 978-1-83910-471-8.

Maher, L. P., Ejaz, A., Nguyen, C. L., & Ferris, G. R. (2022). Forty years of political skills and will in organizations: A review, meta-theoretical framework and directions for future research. *Career Development International, 27*(1), 5–35. https://www.emerald.com/insight/content/doi/10.1108/CDI-07-2021-0191/full/html

Memon, M. A., Ting, H., Ringle, C., Cheah, J.-H., & Muenjohn, N. (2022). Guest editorial: Green human resource management and the implications of culture on its practices in Asia. *International Journal of Manpower, 43*(3), 589–594. https://doi.org/10.1108/IJM-06-2022-598

Mittiga, R. (2021). Political legitimacy, authoritarianism, and climate change. *American Political Science Review*, 11–14. https://www.cambridge.org/core/journals/american-political-science-review/article/abs/political-legitimacy-authoritarianism-and-climate-change/E7391723A7E02FA6D536AC168377D2DE#

Oswick, C. (2010). The social construction of diversity, equality and inclusion: An exploration of academic and public discourses. In G. Healy, G. Kirton, & M. Noon (Eds.), *Equality, inequalities and diversity: From global to local* (pp. 18–36). Palgrave.

Oswick, C. (2012). Discourse analysis and discursive research. In G. Symon & C. Cassell (Eds.), *Qualitative organizational research: Core methods and current challenges* (pp. 473–491). Sage Publishing.

Page, M. J., McKenzie, J. E., Bossuyt, P. M., Boutron, I., Hoffmann, T. C., Mulrow, C. D., & Moher, D. (2021). The PRISMA 2020 statement: An updated guideline for reporting systematic reviews. *International Journal of Surgery, 88*, 105906.

Pearce, J. M. (2016). What if all U.S. coal workers were retrained to work in solar? *Harvard Business Review*, 1–5. 8th August, https://hbr.org/2016/08/what-if-all-u-s-coal-workers-were-retrained-to-work-in-solar

Poschen, P. (2015). *Decent work, green jobs and the sustainable economy* (pp. 1–182). Greenleaf Publishing Limited/International Labour Organization (ILO).

Rodriguez, J. K., Johnstone, S., & Procter, S. (2017). Regulation of work and employment: Advances, tensions and future directions in research in international and comparative HRM. *The International Journal of Human Resource Management, 28*(21), 2957–2982. https://doi.org/10.1080/09585192.2017.1416555

Sandberg, J., & Alvesson, M. (2021). Meanings of theory: Clarifying theory through typification. *Journal of Management Studies, 58*(2), 487–516.

Schuessler, E. S., Lohmeyer, N., & Ashwin, S. (2023). "We Can't compete on human rights": Creating market-protected spaces to institutionalize the emerging logic of responsible management. *Academy of Management Journal, 66*(4) published online 15th August 2023, https://journals.aom.org/doi/10.5465/amj.2020.1614

Shah, J. (2017). 'Executive Order to Create Energy Independence be P****** in the Wind', *LinkedIn,* 28th March, pp. 1–4.

Shudworth, J. (2017a). 'China report on Shijiazhuang', *BBC News,* Accessed 6th January 2017.

Shudworth, J. (2017b). 'China: Renewable Energy Pledge', *BBC News,* Accessed 5th January 2017.

Simpson, A. (2016). 'Businesses no longer have an excuse not to disclose their climate risks', *The Guardian,* 29th December, https://www.theguardian.com/sustainable-business/2016/dec/29/business-climate-risk-disclosure-bloomberg

Singh, D. (2018). 'Green Jobs Horizon Expends in Rapid Climate Changes Era', *ABC Live India,* 1st June, pp. 1–3.

Smith, J. E. H. (2023). 'Stop posting: Elon Musk has killed Twitter – so what should replace it?', *The New Statesman,* 28th March, pp. 1–6.

Stokel-Walker, C. (2023). 'If Elon Musk thinks running Twitter has been painful, he should try using it', *The New Statesman,* 12th April, pp. 1–7.

Sun, P., Doh, J., Rajwani, T., Werner, T., & Luo, X. R. (2023). The Management of Socio-Political Issues and Environments: Toward a research agenda for corporate socio-political engagement. *Journal of Management Studies*, First published 21st September 2023. https://doi.org/10.1111/joms.13002

Valero, A, Li, J., Muller, S., Riom, C., Nguyen-Tien, V., & Draca, M. (2021). 'Policy brief: Are 'green' jobs good jobs? Lessons from the UK and EU to inform labour market transitions', Grantham Research Institute on Climate Change and the Environment/Centre for Climate Change Economics and Policy, *London School of Economics and Political Science,* October, pp. 1–8, https://www.lse.ac.uk/granthaminstitute/publication/are-green-jobs-good-jobs-how-lessons-from-the-experience-to-date-can-inform-labour-market-transitions-of-the-future/

Vincent, S., Bamber, G. J., Delbridge, R., Doellgast, V., Grady, J., & Grugulis, I. (2020). Special issue: Situating human resource management practices in their political and economic context. *Human Resource Management Journal, 30*(4), i–iv. 461-623 (November), https://onlinelibrary.wiley.com/toc/17488583/2020/30/4

Yin, R. (1994). *Case study research: Design and methods* (2nd ed., pp. 1–170). Sage Publications Ltd. ISBN 0-8039-2663-0.

3

Britain in Decline: Politics, Green Job Trends and Issues Arising

Abstract I reflect on the notion on the UK in decline and use the example of Britishvolt to support it. I then critically discuss UK politics, green job opportunities and trends, before offering a critical view of them, and UK green job developments and issues emerging.

Keywords UK • Decline • Politics • Britishvolt • Trends • Issues • Opportunities • Developments • Critical

Introduction: UK in Decline

Britain faces several well-known issues contributing to its current 2020s-era economic and political problems, namely: management, inherited wealth, lack of competition and skills shortages, as UK social divisions have produced 'Broken Britain' and the return of "private affluence and public squalor" (Lansley, 2023, p. 1). Here, economists Van Reenen and Bloom's World Management Survey (WMS) found that 'over half of the productivity gap between the UK and the US could be due to poor management', and UK family firms are 'more likely to practice primogeniture' (where property passes down to the eldest son), 'whereas a leading

D. W. S. Renwick, *Work, Politics and the Green Industrial Revolution*,
https://doi.org/10.1007/978-3-031-53067-8_3

predictor of good management is that firms are exposed to strong competition'. Such low competition 'led to the poor management of British businesses during the post-war industrial decline'[1] (in The Economist, 2023, p. 22) which continues today. As one now prominent pro-green UK conservative politician states:

> Net zero is not a crusade, it is the only proportionate and pragmatic way in which the UK can avoid higher costs over time and establish the industries of the future, rather than be stranded in the past. Net zero is not about ideology, it is about delivering jobs and growth and the UK leading the change, or following, missing out in turn on the investment and jobs that could have been ours for the taking. (Chris Skidmore MP in Crerar and Horton, 2023, p. 8)

Britishvolt

One example of poor UK management occurred in the green policies British conservative politicians enacted at Britishvolt—which was a start-up formed in 2019 to build batteries for electric cars based in Blyth, North-East England. Then, Britishvolt was established with large UK government subsidies promised, yet three years later it was considering financial administration and closure (Edgerton, 2022, p. 1). This circumstance partly happened because the British government did not award Britishvolt a promised £100 million advance for tooling equipment, but instead produced images of a new UK Gigafactory when none existed and publicized £1.7–3.8 billion of investment when the real figure was in the tens of millions (Edgerton, 2022, pp. 1–2). A key problem emerging was British hubris and 'boosterism' under successive conservative governments led by PMs David Cameron, Theresa May, Boris Johnson and

[1] Essentially, I break with the UK-based, conservative-led argument that Britain is not in decline advanced by UK Chancellor Jeremy Hunt, who criticizes 'insidious decline-ism' (Hunt, 2023, p. 4), and the interlocking case that UK problems stem from a low worker productivity, e.g. Ashworth-Hayes (2023a, p. 14; 2023b, p. 21), Jacobs (2023, p. 14), and Timpson (2023, p. 22). Instead, I use economic growth and social inequality outcome data from Sikka (2022) to argue that UK problems arise from unequal and inconsistent British government policy, poor organizational leadership and daily managerial incompetence.

Rishi Sunak. Here, for over 40 years such UK governments pursued policies based on notions of 'British innovative genius that just needs realizing by creating start-ups', and as the UK now recently claimed leadership in battery technology, wanted to lead it with a British firm, to chime with new UK political independence under Brexit (Edgerton, 2022, p. 2). However, as Britain has no world leaders in the battery sector, and only represents roughly 2–3% of global research and development and manufacturing in it, more realism seems needed in their talk about it (Edgerton, 2022, p. 3). This occurred especially as the UK government later refused Britishvolt's request for a £30 million emergency loan (Keating, 2022, pp. 1–2), and the lack of such monies partly led Britishvolt to collapse, alongside it 'failing to secure a single order' (Marlow, 2023, p. 20). Thus, Britain is now seen to have failed to build up a substantial homegrown battery industry (Mustoe, 2023a, p. 18).

Moreover, the Australian-based start-up Recharge Industries which took over Britishvolt in early 2023 (Evans-Pritchard, 2023a, p. 24) missed a final, key payment and fuelled rumours of no buy-out and no future for it. Thus, it seems obvious why new UK PM Rishi Sunak seemingly handed Tata £500 million in subsidies to build their new £4 billion plant in Somerset making batteries for Jaguar Land Rover (JLR) (Mustoe, 2023b, p. 21). This is because 'without enough giga factories, Britain could lose out on many of the nearly 100,000 jobs linked to making cars and internal combustion engines', with industry experts estimating another three such gigafactories being needed for the UK to replace such car and engine output (Jolly, 2023, p. 51). Some commentators advise that Britain should avoid the European rush to build 'low-margin' gigafactories and instead focus on solid-state, sodium-iron and lithium-sulphur technologies because with Britishvolt the UK got caught 'botching it last time' (Evans-Pritchard, 2023a, p. 24), as 'somewhere between 2017 and early 2023, the British government dropped the ball' (Evans-Pritchard, 2023b, p. 23). A key question remains if the former Britishvolt site will stimulate further, related, new investment and UK-based green jobs or become a one-off living in the shadow of the British government's failure to do so.

UK Politics and Green Job Opportunities

Within other sectors such as steelmaking, while new British PM Rishi Sunak has called for 'bold, persistent innovation to turn the tide on UK economic woes', it is other countries that are the leading, global steelmakers, engineering and mining firms, and undertaking trials of carbon capture utilization and storage (CCUS) technology—as seen at ArcelorMittal's steel plants in Belgium and North America (Holder, 2022, pp. 1–2). There, firms are using hydrogen and electrification in steelmaking to replace fossil fuels, and CCUS to potentially decarbonize current CO_2-intensive blast furnaces. However, even such technology leaders do not yet have fully operational CCUS systems in place (Holder, 2022, p. 3). Such developments help explain why some British-based critics argue that PM Sunak has 'consistently failed to address the climate crisis which is the other side of the energy coin' (White, 2023, p. 3), while his political opponents in the UK Labour Party have accused him of 'economic illiteracy' and lacking seriousness after he said Britain would become energy independent but at the same time made it harder to use onshore wind (Walker, 2022a pp. 1–2). Sunak has come under pressure from his conservative party activists who do not like wind turbines installed in rural UK locations and opposition politicians who see rising energy bills as frustrating them because onshore wind sites could be working and providing low-cost clean power for British billpayers in roughly one year (Walker, 2022b, pp. 1–2). PM Sunak responded to such criticisms by enhancing his green rhetoric, as he said: "We can turn our struggle against climate change into a global mission for new jobs and clean growth, and we will now triple our funding on adaptation to £1.5 billion by 2025", yet UK government ministers also suggest there will be no change regarding climate issues under Sunak (Maidment and Penna, 2022, pp. 5–6). Here, Rishi Sunak initially blocked new onshore wind farms upon taking office but was then forced to back down and announce a consultation on it in December 2022 (Riley-Smith, 2023, p. 1).

One example of PM Sunak's supposed potential greenness lies in his creating a new government Department for Energy Security and Net Zero, thereby giving net zero Cabinet level representation for the first time (Kapadia, 2023, p. 18). He is also considering 'a plan to charge motorists for every mile they drive to reduce a £40bn fuel tax gap left by Britain's early move to electric cars' and is reported as being 'very interested' in a pay-as-you-drive system too (Robinson and Maidment, 2020, p. 1). Here, younger Britons are contributing to declining UK car ownership, as nearly one-third in England now use electric scooters and bicycles, while the banning of the sale of new petrol and diesel-owned cars from 2040 is predicted to create over 30,000 new jobs and a £4.2 billion boost to the British economy (Robinson and Maidment, 2020, pp. 6–8). Yet PM Sunak still maintains the pro-fossil fuel status quo too, as one University College London (UCL) report states the UK government's backing for the Sizewell C nuclear plant is 'increasingly difficult to justify' because their bill for it will probably be over £30 billion, and as the British government 'didn't even fully disclose what its "historic" £679 million spend on it is for' (Dorfman, 2022, p. 1). Further, while a UK government review of net zero is ongoing in 2023, groups such as Greenpeace state that British onshore wind farms are effectively 'banned', UK-based water companies can dump hundreds of thousands of tonnes of raw sewage into rivers and seas including the Thames (Greenpeace UK, 2022b, p. 1; 2022c, p. 1). Moreover, Sunak seems wholly uncomfortable when talking about the climate emergency, as he did not mention it in his key speech as PM in January 2022 (Goodman, 2023, p. 21), and he instead approved up to 130 new oil and gas licences in the North Sea (Greenpeace UK, 2023, p. 1), Britain's first new coal mine in 30 years (Greenpeace UK, 2022d, p. 1), and up to £20 billion funding in July 2023 for carbon capture CCUS projects too (Oliver, 2023, p. 17).

One rationale Sunak's government gave regarding their new UK coal mine approval is that they expect it will generate 500 jobs, but the Local Government Association (LGA) says there is conversely potential for 6000 'green jobs in Cumbria by 2030' without it (in Sharma, 2022, p. 1). Here, experts say they cannot find one steel firm which wants new Cumbrian coal (Harrabin, 2022a, p. 1), that this mine approval illustrates 'the power of legacy industries over our policies' (Monbiot, 2022a, p. 1), and that the UN climate change chief has criticized it too (Vaughan, 2023, p. 1). Moreover, under Sunak Britain's long-term air pollution targets were published six weeks late and seen as 'outrageously unambitious'

(Mums For Lungs, 2023, p. 1), and he was personally criticized for moving in the wrong direction by encouraging more domestic UK flights that increase emissions and had to be 'handed a get out of jail free card' as Rolls-Royce 'pledged to decarbonize planes' too (Paul, 2022, pp. 1,7,9).

One issue that the new Sunak government currently seems to miss is the complexity involved in any new British green job creation. Here, analysts detail sets of 'green' and 'brown' jobs, where green jobs concern 'activities consistent with delivering the net zero transition', while others requiring significant change to be consistent with net zero are classified as 'brown changers' (Broome et al. 2022, p. 3). Core green task occupations 'represent 13% of UK employment, or 4.3 million workers, while 40% of workers in both green and brown jobs work in construction, manufacturing, agriculture, or the energy sector, compared to 20% of workers overall. Here, green jobs (but not brown ones) are 'overrepresented' in the highly paid, management, professional, scientific, and technical sectors, especially in parts of the south-east, London and Malton in Yorkshire', and issues emerging include the British government's inability to tackle sector, occupation, regional and pay differences (Broome et al., 2022, pp. 4–5).

Further, complications for Sunak's government arise due to brown jobs being more located in process and plant operative occupations, and men being more likely to work in green and brown jobs rather than women (Broome et al., 2022, p. 6). But it is not clear if and how Sunak intends to level out/up such UK differences. Here, experts argue that meeting British emissions targets requires much green job growth in the next ten years, as industry estimates are that the UK may employ roughly 70,000 workers in offshore wind (up from 26,000 in 2021), an additional 260,000 energy workers being needed by 2050, and 60,000 more workers required to install heat pumps over the next seven years too (Broome et al., 2022, p. 7). However, relatively few people have moved from brown to green jobs, while 'the share of brown job workers moving to British-based green jobs has been low (2.8%), with movers tending to be more educated, and coming from brown jobs with a high content of analytical tasks', meaning further transitions require more work from business, government and staff, and 'such efforts are likely to vary across place, sectors, and demographic groups' (Broome et al., 2022, p. 9). Yet little is heard from the Sunak regime on these changes, and on how he intends to tackle the issues emerging—of vacancy filling, training, skills and transitions. I now chart some British-based government-led green job trends before critiquing them.

UK Green Job Trends

Britain's progress towards an electric vehicle (EV) revolution includes at least 90,000 fully electric and plug-in hybrid cars on UK roads, with less than 1% of its 33 million cars as plug-ins (Duell, 2020, p. 11), one battery car being registered every 60 s (Murray, 2023, p. 1), EVs making up over one-third of all new cars produced and the proportion of fully electric new cars sold could be 1 in 12 by 2030 (Vaughan, 2017, p. 1). Issues around an upturn in British electric car use include the many new charging points required to plug them in (Duell, 2020, p. 11), relevant job vacancies arising to install them, and the future of the new Coventry factory building the new electric black cab (TX5) taxi (Vaughan, 2017, pp. 1, 3), as they all show the private sector is needed to make Boris Johnson's original green jobs plan work fully (Jones, 2020, p. 1). Here, a report by the London School of Economics and Political Science (LSE) reveals several important areas where UK government investment could create jobs, namely, energy efficiency in buildings; natural capital projects; active travel equipment and infrastructure (bicycles and cycle lanes), renewable power generation, distribution and storage; electric vehicle production and charging infrastructure; CCUS and blue and green hydrogen production; and manufacturing parts for wind turbines. They recommend that efforts to create such British-based jobs in the short term should be complemented by 'investments, policies and incentives for building innovation and skills to meet these longer-term challenges' (Unsworth, 2020, p. 2). Yet fault lines exist in Boris Johnson's original drive to increase UK hydrogen production and replace gas boilers by 2023, as only two such prototypes exist, one-fifth of the British gas network needs re-laying and all engineers to be retrained. Additionally, no single hydrogen boiler is yet for sale in the UK, the Home Builders Federation (HBF) says the 2023 gas boiler ban deadline produces huge challenges for developers, suppliers and energy firms regarding skills, design and infrastructure, and targets to install 600,000 UK heat pumps each year by 2028 are unlikely to be delivered (Duell, 2020, pp. 1–4). Thus, if Boris Johnson's green agenda is to work, it needs to 'deliver UK jobs and economic growth' too (Warner, 2020, p. 1).

British environmental plans offer a mixed bag towards creating a post-COVID-19 green jobs boom, as only five such jobs 'on average are created per £1 million spent' (Ingham, 2021, p. 1). A particular problem lies in home installations, where some UK installers in Lancashire spent time wondering how not to make redundancies, after many years when ministers said they wished to rapidly increase insulation in British homes, only to later discard such plans. Previously, funding to meet home insulation targets came from the UK government's Green Homes Grant (GHG), but this suffered problems including payment delays, meaning the Department of Business, Energy and Industrial Strategy (BEIS) had to extend the scheme until March 2022 (Ingham, 2021, p. 5). Michael Pollitt at Cambridge University adds that 'while 35 jobs are created in the broader recycling industry for every £1 million invested, only 10 jobs are created by opening a Gigafactory (battery factory)', whereas the wind-farm sector generates two jobs in comparison (in Ingham, 2021, p. 7). Therefore, government delays and short-termism are issues Britain faces regarding home insulation upgrades which continue today (see Skidmore, 2023, p. 1).

A related but different issue concerns the UK government's optimism about their estimates of green job growth, which seem quite high. For example, one Boris Johnson press release relating to the Humber Teesside and the British energy industry argued for the 'next generation of off-shore wind turbines whilst creating 6,000 new green jobs in the process [with] GE Renewable' (Johnson, 2021, p. 1), while the Onward think tank estimated his eco-plan could 'create a staggering 1.7 million new jobs', where half the new eco-jobs 'in his green revolution can go to the North and the Midlands' (Clark, 2021, p. 1). More modestly, the UK was also reported to 'create and protect 1,340 jobs with a "major" new green energy project across the Humber region', thanks to a £266 million investment by offshore wind manufacturers Siemens Gamesa and GRI Renewable Industries (Hoare, 2021, p. 2).

The optimistic narrative above was continued in the British government press release titled: 'Prime Minister's Ten-Point Plan[2] kickstarts green investment boom' (GOV.UK, 2021a, p. 1), which reported large numbers of UK green jobs being created. Here, Britain 'has attracted almost £6 billion of green investment, and figures show 56,000 high-quality green jobs secured and created across the UK since November 2020', offshore wind supporting around 3600 jobs across the Humber and North-East, approximately 56,000 green jobs across the UK economy 'already online or in the pipeline over the next decade', and examples 'in electric vehicle manufacturing in Sunderland, hydrogen in Teesside and offshore wind in Northumberland, Yorkshire and Humber' (GOV. UK, 2021a, pp. 2–3). The former UK Business and Energy Secretary Kwasi Kwarteng said: 'we are laying the foundations for a new era of green jobs and decades of economic growth' (GOV.UK, 2021a, p. 4), and the British government added: "The PM's Ten Point Plan will mobilize £12 billions of government investment over the next decade to create and support up to 250,000 highly skilled green jobs – and unlock three times as much private sector investment by 2030, with which UK government investment together approximately totaled roughly over £4.5 billion" (see GOV.UK, 2021a, pp. 5–6 for a financial breakdown). The British government added that they have 'secured £9 billion of deals that will create more than 30,000 jobs across the UK' (Trevelyan, 2021, p. 1), and 'almost £10 billion that will drive growth and high-skilled jobs across the country, [plus] Iberdrola's £6 billion planned investment in offshore wind will create 7,000 high-skilled jobs' too (GOV.UK, 2021b, p. 1). They additionally see progress being made regarding British green job growth, now requiring all new UK buildings to install EV charge-points which may stimulate green jobs (Department for Transport (DfT) 2021, p. 1), and distilleries will use hydrogen and biogas in production processes, which supports more green jobs due to over £11 million in new funding (Business, Environment and Industrial Strategy (BEIS) 2021a, p. 1).

---

[2] On chronology, the UK government released Boris Johnson's *Ten Point Plan* first (late 2020), the *GJT* report (July 2021), this (2021a) file above and then their *Net Zero Strategy* (Dobbins, 2022, p. 34).

Chris Skidmore MP (who became Chair of the British government's new Net Zero Review, NZR group under PMs Truss and Sunak in 2022–2023) states that Humber's offshore wind industry has been developed by investment from Siemens and other international companies (Northern Powerhouse Partnership (NPP) 2022, p. 1), while East Anglia has almost half the UK's installed capacity of offshore wind which brings jobs and investment (in Hubbard, 2022, p. 1). Some 20,000 social housing property upgrades can back 9000 jobs (Business, Energy and Industrial Strategy (BEIS) 2022a, p. 1), and making turbines, EVs, and fitting insulation and heat pumps can become skilled jobs too (Beard, 2022, p. 1). Additionally, more young people now focus on climate-related work and are seeking green jobs (Kantor, 2021, pp. 1–2), including Kent County Council graduate opportunities which focus on flood and water management, highways, governance, law and democracy (Promoting People, 2022, p. 1). The British government add that new tech 'will generate clean hydrogen from biomass' and create new green jobs (Business, Energy and Industrial Strategy (BEIS) 2022a, p. 1), lower green levies are stimulating investment and jobs 'into post-industrial regions' (Skidmore, 2022, p. 1), and a new energy security strategy could create more than 40,000 extra UK-based green jobs too (Business, Energy and Industrial Strategy (BEIS) 2022b, p. 1). Indeed, domestic hydrogen production looks set to 'create tens of thousands of jobs and produce a super fuel that could power British industry' (Kwarteng, 2022a, p. 1), while replacing aviation with train journeys could mean total UK employment 'increase by 280,000 jobs in low-emission travel, domestic tourism, and R&D' (Possible, 2022, p. 1).

Moreover, Scottish Power has announced 1000 new jobs on 16 renewable energy projects (Kwarteng, 2022b, p. 1), work clearing the Sinter Plant on Teesside should produce a world-leading 1 GW power station with CCUS, 'generating 5,000 construction and hundreds of operational jobs' (Houchen, 2022, p. 1), and if the UK develops 30 GW of onshore wind by 2030, doing so could create 27,000 high-quality jobs and 30,000 more in operations and maintenance (Bennett 2022, p. 1). The latest PwC Green Jobs Barometer reveals the number of British green jobs growing at approximately four times the overall employment market rate (Murray, 2022a, p. 2), and National Grid estimates that the UK needs to generate around 400,000 green jobs to transition to a net zero economy by 2050 (Jeraj, 2022, p. 2). Further, two Scottish green freeports in Cromarty Firth and the Forth look set to create over 75,000 new jobs (Hardman, 2023, p. 1), and one huge wind farm planned in Scotland could create 6000 full-time jobs (Hudema, 2023, p. 1), and investment of £32.9 million will create a network enhancing high streets which may see up to 1300 new green jobs being created (Active Travel England, 2023, p. 1).

Lastly in this section, Jaguar Land Rover is advancing electric car production and securing British-based jobs (McLoughlin, 2021, p. 1), BP, SeAH and Tees Valley Lithium could become job creators at Europe's largest brownfield site at Teesworks, and 6000 jobs are planned at the NZT Power plant (forming natural-gas with built-in carbon capture). H2Teesside wants to be a large UK blue hydrogen production facility, SeAH Wind factory is producing legs for wind turbines and alfanar wants to make sustainable aviation fuel (Hughes, 2023, pp. 3–4). Further, British-based green jobs are expected from aviation leader Willis Lease Finance Corporation, the Jet Centre at Teesside airport and a Rolls Royce factory making parts for their fleet of clean energy power stations. The UK's EV car revolution may also support up to 1000 jobs at Tees Valley Lithium, while in Sunderland 'the Quantafuel recycling plant and Wastefront's tyre recycling one look set to bring another 100 jobs each' (Hughes, 2023, pp. 3–4).

Therefore, if viewed comparatively, it is possible to see how future British green job creation figures (detailed above) may fit the original Boris Johnson era estimate of up to 250,000 new UK-based green jobs being created (Bhate, 2023, p. 1). Moreover, a lot is going on in the initiatives above to create new, British-based green jobs, and government ministers and business groups can and do take credit for it. However, the UK plans here also look slow and ponderous when compared to the much more expansive and lither EU and US ones (detailed in Chap. 1), especially regarding what happened to Britishvolt too (detailed above). So, while the British government and relevant industry figures on actual and predicted claims regarding UK green job creation are clear (above), they do not always show the downside of such initiatives—which I now explore.

A Critical View

Critics argue that Britain's net zero plan lacks ambition and funding, as while the government argues that their strategy would create 440,000 jobs, the Treasury also warns that taxes may need to increase to fund such changes. This is because analysts reveal around £26 billion of funding going into UK green plans would be drawn from the public sector over the 2021–2025 spending review period, with over £60 billion needed from the private sector too. Additionally, the Net Zero Review document published by the Treasury reveals that the British government is likely to lose billions in fossil fuel tax revenue (Harvey and Elgot, 2021, pp. 1–3). Thus, the up to 440,000 new job plans detailed earlier were criticized because the UK government were not keen to clarify how many may arise from their strategy rather than through jobs which transition to net zero, and Greenpeace argue that the British government lacks clear plans to scale-up renewable energy, invest more in public transport and stop new oil and gas licences (in Harvey and Elgot, 2021, pp. 3–4).

Academics such as Kevin Anderson at Manchester University argue that the UK's green revolution 'risks being tripped up by lack of green jobs' because the numbers involved 'reveal a story of subterfuge, delusion, offsetting and piecemeal policies' (in McCulloch, 2021, pp. 1–3). Further, British employers in the green and clean sectors need to consider flexible working and evidence-based training (Ormsby, 2021, pp. 1–2), UK Members of Parliament (MPs) state that the British government's green jobs plan 'needs more clarity', and the UK Environment Audit Committee (EAC) argue that it will fall away without such precision, as the government are 'yet to define what a "green job" is', how they will assess the demand and skills needed and where such jobs are to be created (May, 2021, p. 2). Here, the EAC recommend that environmental sustainability is included in all British national curriculum, A-level, apprenticeship and T-level courses and that the UK government should detail a programme stimulating the development of green skills within the construction sector to generate capacity and national careers strategy plus employment placement schemes in green sectors too (Faragher, 2021, pp. 1–2).

Additional criticisms of the British government's approach to green jobs are that ministers need to speed up to reach their targets (Kennedy, 2021, p. 1) and that typical salaries for UK-based green vacancies vary, as the median pay (of £18,270) for recycling workers is much lower than the average full-time annual salary (of £31,000). Indeed, only environmental managers/sustainability consultants (earning £37,500 and £42,500 respectively) are seemingly well-paid in comparison, while slow job creation and few green roles exist, meaning British policymakers need to examine new training programmes to help employees gain necessary skills (Kennedy, 2021, pp. 1, 4, 5). With only 0.2% of UK postings classified as green jobs,[3] and roughly 0.12% of all such jobs classified as green in 2019 by the Office for National Statistics (ONS), the EAC argue that a detailed plan is needed to deliver such green job growth ambitions (Bernard, 2021, pp. 1–3), while the British government frustratingly asks: 'what is a green job?' (Business, Environment and Industrial Strategy (BEIS) 2021b, p. 1). At the same time, overseas observers state that the UK is not proactively decarbonizing the steel sector, or investing in new green jobs in coal in Whitehaven, Cumbria (Littlecott, 2021, p. 1), that decarbonizing British railways will support much green job generation (Rail Delivery Group (RDG) 2021, p. 1) and needs to focus on delivering ambitious climate action overall (Guterres, 2021, p. 1).

Further, some critics see extra work being needed to help increase UK green job growth, because the green economy only accounted for '1.2% of all advertised roles in the year to July 2021, or just 124,600 new jobs' (May 2021, p. 1). Here, the British Trades Union Congress (TUC) warn that up to 660,000 jobs could be risked if the UK fails to achieve its net zero target as quickly as other countries, and PwC chairman Kevin Ellis states that "evidence is needed on the level and distribution of these opportunities" (in May 2021, pp. 1–2). Other relevant yet disparate issues include trenchant skills shortages which might see green opportunities being exhausted (Business Green, 2022, p. 1), some staff at the British Environment Agency feeling less able to complete their jobs as staff morale is too low (Salvidge, 2022, p. 1), the national curriculum not

[3] 'Clean industries already support 410,000 jobs across the UK' (Business, Environment and Industrial Strategy (BEIS) 2021a, p. 1).

incorporating sustainability, UK government departments not centrally coordinating green jobs policies (Harrabin, 2022b, p. 1) and over 30,000 jobs being risked if the insulation levy is removed from fuel bills (Carrington 2022, p. 1). More recently, rising living costs have made achieving net zero politically trickier, as the British government must demonstrate the broader goal of economic benefits arising from creating green jobs, and funding clean technology via general taxation too (Buckland, 2022, p. 1).

UK Green Job Developments and Issues Emerging

Commentators state that British Office for National Statistics (ONS) figures detail UK jobs being down in energy-efficient products manufacturing by over a quarter, in onshore wind by nearly one-third and in solar by over 40%, while low-emission vehicles and infrastructure jobs have more than doubled (upwards to 19,100 in 2020). Here, the general British trend is downward in numbers of green jobs, and only a quarter of workers surveyed think the UK government's current climate plans will generate more new jobs in their area, while TUC General Secretary Frances O'Grady states that: "we can create 300,000 jobs in retrofits, it's time for ministers to get on with it" (Trades Union Congress (TUC) 2022a, pp. 2–3). The overall number of people working in UK green jobs has shrunk by 12%, which led Roz Foyer at the Scottish TUC to say that: "on new green jobs, the truth is we have fewer than we did 8 years ago, and the number continues to fall" (Trades Union Congress (TUC) 2022b, p. 1). Indeed, records show large differences in green occupational employment job numbers, as previously available figures in Scotland reveal they stood at 745,165 such jobs in 2014, which is a higher figure than the Scottish government's narrower focus in the renewables sector of 43,500 green jobs (Scottish Government, 2017, in Sofroniou and Anderson, 2021, p. 34).

Critics fear that the UK sustainability 'skills crunch' is getting worse (Murray, 2022b, p. 2), and highlight that while Britishvolt's Gigafactory in Northumberland was supposed to create 3000 direct jobs, and 5000 more among the supply chain (Jefferson MFG, 2022a, p. 1), neither happened. Further, observers point to 2016, when British conservative political leaders George Osborne and David Cameron 'killed off 12,000 solar jobs in a year, and now Sunak and Johnson want us to get excited about the prospect of 10,000 "new" solar jobs by 2028. Unserious energy policy in serious times' (Bell, 2022, p. 1). From such developments, we can see that the UK's green economy has not increased much since 2014 and that the new Energy Strategy says little about worker training too (Greenpeace UK, 2022a, p. 1). Here, critics add that Rishi Sunak, the Environment Secretary, Transport Secretary and many UK government advisors do not comprehend that they need to do more than just say that they are committed to net zero (Murray, 2022c, p. 1), that the UK is working with a fifth Energy and Climate Minister in less than two years, and a sixth Business Secretary within four years (Murray, 2022d, p. 1). Indeed, the current 2023 Environment Secretary Therese Coffey served before in Defra as a junior minister, where 'to put it very politely, she is not remembered with fondness' (Monbiot, 2022b, p. 1), and was seen as 'complacent' when handling water supply shortages (Horton, 2023, p. 1).

Unfortunately, British green jobs do not currently all seem like good jobs, as while the UK Net Zero Strategy predicted 440,000 jobs by 2030, and Boris Johnson's original Ten Point Plan for a Green Industrial Revolution pledged 60,000 jobs too, the decarbonization process itself creates new winners and losers. These include communities who rely on carbon-heavy work, as one in five British jobs could be impacted geographically from 'airport towns like Hounslow to oil and gas jobs in Aberdeen', meaning possible 'jobs lost, local economies disrupted, and communities broken' (Atkins, 2021, pp. 2,4). Here, previous UK conservative government policy and action does not bode well for the future of such impacted workers, because while the last enabled decline of the British coal mining industry saw some schemes focused on 'retraining redundant miners, encouraging entrepreneurialism, and creating coalfield "enterprise zones", none proved successful', meaning that the UK now needs a new, proactive green industrial strategy to soften the impacts of other policies (Atkins, 2021, p. 6).

Moreover, commentators talk of lower wages and losing skills, as decarbonization does not always mean new green job creation and any such jobs could be low-paid and insecure, because British offshore wind projects have 'used irregular migrant labour, paying substantially below the minimum wage, and demanding extensive working hours' (Atkins, 2021, p. 6). Further, it is unclear how skills and training will be transferred from carbon-heavy industries into the renewables sector, although some skills can be transferred, e.g. plumbers and pipefitters in the gas sector moving to green hydrogen, and oil rig operatives working in offshore wind. But while the oil and gas sector require two qualifications, no UK training passport system exists to recognize the other qualification, and impacted workers are concerned about their job security, as community-led small-scale rooftop solar work tends to be ad hoc and localized meaning lower job security (Atkins, 2021, pp. 2, 7, 9). Hence The Society of Motor Manufacturers and Traders (SMMT) warn that even successful moves to EVs will see thousands fewer jobs being available and workers retraining, and only 226,000 jobs created in new battery production (McLoughlin, 2021, p. 4). Indeed, the British government's BEIS department and the Institute for Apprenticeships & Technical Education (IATE) see a need for assimilating sustainability into developing new apprenticeships and technical qualifications too (Institute for Apprenticeships and Technical Education (IATE) 2023, p. 1).

A challenge lies in the UK government needing to understand the scale of change required in adapting to world electricity grid expansion and greening, which may roughly cost $1.1 trillion a year between now and 2050, with existing projects needing to be speeded up, and some strong objections to building made on environmental grounds, e.g. to maintain skylines or preserve woodlands (The Economist, 2023, p. 7). Further, some commentators see that the impact of the net zero by 2050 transition could affect Britain unequally, as a study by the Place-Based Climate Action Network (PCAN) found roughly three million jobs needing much change and upskilling, and another three million jobs requiring increased demand, while upskilling workers (25%) may be needed to adapt, and without it such jobs risk disappearing (UK Research and Innovation (UKRI) 2023, p. 2). Additionally, this transition may become unequal

(Draca et al. 2021, p. 1), as women and younger workers may need specific policies to redress 'regional and demographic imbalances' too (Valero et al., 2021, p. 1).

A cross-parliament group of British politicians tabled a Green New Deal Bill which included calls for an action plan to create 'millions of good green jobs across the UK' and ensure they 'are good quality as defined by the Good Work Plan, regardless of skills base' (in Dobbins, 2022, p. 34). Yet while the GJT offers one route for adaptation to satisfy such needs more permanently, it also raises concerns from many relevant stakeholders 'that the Good Work agenda is falling by the wayside as a result of changing government policy priorities' (Dobbins, 2022, p. 35). For example, in oil and gas, many workers desire green reskilling in North-East Scotland, but the move from fossil fuels to green energy is not so smooth there (Thomas, 2022, p. 1).

Such findings reinforce Professor Linda Clarke's case that a lot of green job transitions 'do not alter social relations of power and employment relationships' (in Dundon, 2022, p. 1), and that to ensure a fair transition, aviation workers are asking for a 'right to retrain scheme, and a frequent flying levy to discourage flying' (Possible, 2022, p. 1). Further, Scottish trade union leaders are calling for an end to 'broken promises' to workers, and the start of a 'renewables jobs revolution' (Scottish Trades Union Congress (STUC) 2022, p. 1), as quite predictably, jobs in fossil fuels are not now seen as lifelong employment, because less need may exist for them in future (Ralston, 2021, p. 1).

In other areas, approximately 6.3 million British jobs (roughly one in five) may become impacted (positively or negatively) by the green economy transition, with sectors like construction, manufacturing and transport probably most affected (London School of Economics and Political Science (LSE), 2021, pp. 1–3). Evidence supporting this criticism comes from offshore oil and gas workers spending over £2000 of their own money on retraining, but still fearing redundancy (Holland, 2021, p. 1). Indeed, social dialogue, tripartism and collective bargaining also seem needed in designing climate transition policies (Singh, 2018, p. 3), as a survey of over 600 offshore UK-based oil and gas workers found that hundreds face redundancy and cannot gain renewables jobs, but many of those impacted want to remain anonymous, over fears they may become grey listed in the industry for raising any issues. One worker said: "I haven't seen any help at all, there's nothing being done, and everyone's really angry" (in Holland, 2021, pp. 6–10). Further, the TUC argue that 'up to 660,000 jobs, many in Britain's political "red wall" industrial heartlands are at risk unless the UK government speeds up green investment', and that they should fully commit to the GJT plans, because unions want a government-led £85 billion green recovery plan to generate 1.24 million green jobs, and avoid 260,000 manufacturing jobs and over 400,000 in the supply chain (660,000 overall) moving offshore. Here, TUC figures reveal that the British Treasury 'is investing £180 per person on the green recovery and jobs over the next decade compared to President Biden planning £2,960 per person', while Alan Coombs at Port Talbot Steelworks said: "companies overseas are already setting target dates for green steel. But the UK isn't even putting our toe in the water" (in Craig, 2021, pp. 1–11). Ergo, PM Rishi Sunak is rumoured to be 'in advanced talks with Tata Steel for a £500 million aid package' to secure Port Talbot steel and build electric arc furnaces, yet 3000 jobs could be lost in the long run there even after his state aid package arrives too (Almeida, 2023, p. 9).

Thus, some observers conclude that the British government must do more to develop the green factories of the future or risk losing industrial jobs, yet also acknowledge that such large changes mean disruption, especially regarding the political need to keep trust in forgotten UK communities in the North and the Midlands (i.e. car manufacturers, steel producers and high-carbon industries). They also acknowledge that such dislocations also bring opportunities, in electric vehicles (Nissan, Sunderland), offshore wind turbine manufacturing (Siemens, Humberside) and redeveloping the Teesworks (Birkett et al. 2022, pp. 1–3). Nonetheless, many dangers exist, as the British government's time for action is roughly five years to secure investment in manufacturing green technology factories (electric vehicle batteries, offshore wind turbines and steel production), and if not such factories may go to competitors in Europe, the US and Asia, while any job losses could impact heavily on industrial communities still suffering from a decline in shipbuilding and coal mining (Birkett et al. 2022, p. 8). Recently, evidence is that the UK is falling behind in securing offshore wind manufacturing jobs, as firms there prefer to build capacity in Europe where governments offer more state-led financial support for capital and port infrastructure, and a more state-led, interventionist approach only occurred in Britain under the Theresa May and Boris Johnson conservative UK governments (Birkett et al., 2022, p. 8), which implies that less support is now happening under the new Sunak administration. Overall, as one commentator states:

> Many see something quintessentially British in the country's ongoing slide – an unsensational story of disgruntlement, dysfunction, and decay … yet there's a widespread complacency among the elite about the possibility that Britain could be heading for a genuine crisis. The question is how on earth we get out of this trap. But this makes one very hasty assumption: that anyone knows in detail what needs to be done. (Jacobs, 2023, p. 14)

I now detail one potential answer on what needs doing above, in outlining and critiquing the British Green Jobs Taskforce (GJT).

References

Active Travel England. (2023). *Twitter/X,* 2nd January, p. 1.

Almeida, L. (2023). 'Tata Steel close to £500m state aid deal to help with going green', *The Daily Telegraph,* Sunday 3rd September, p. 9.

Ashworth-Hayes, S. (2023a). 'The UK is a poor country that is determined to get even poorer', *The Daily Telegraph,* Wednesday 26th July, p. 14.

Ashworth-Hayes, S. (2023b). 'Workless Britain is still on permanent furlough', *The Daily Telegraph,* Thursday 24th August, p. 21.

Atkins, E. (2021). 'Why green jobs aren't good jobs yet', *Bristol Inclusive Economy Initiative,* 3rd November, pp. 1–10.

Beard, J. (2022). *Twitter/X,* 18th February, p. 1.

Bell, A. (2022). *Twitter/X,* 7th April, p. 1.

Bennett, N. (2022). *Twitter/X,* 23rd September, p. 1.

Bernard, D. (2021). 'Growth in green jobs market remains sluggish', *HR Magazine,* 27th October, pp. 1–3.

Bhate, R. M. (2023). 'The fastest growing and declining jobs in the next decade', LinkedIn, 3rd January, p. 1.

Birkett, E., Luke, A., & Bunt, P. (2022). 'Green jobs, red wall: why green industrial jobs are critical to levelling up', *Onward UK,* 16th December, pp. 1–10, https://www.ukonward.com/reports/green-jobs-red-wall/

Broome, M., Cellini, S., Henehan, K., McCurdy, C., Riom, C., Valero, A., & Ventura, G. (2022). 'Net zero transition to mean significant change for 1.3 million workers', *LSE Blog,* 30th June, pp. 1–12, https://blogs.lse.ac.uk/businessreview/2022/06/30/net-zero-transition-to-mean-significant-change-for-1-3-million-workers/

Buckland, J. (2022). *Twitter/X,* 5th January, p. 1.

Business, Energy and Industrial Strategy (BEIS). (2022a). *Twitter/X,* 12th January, p. 1.

Business, Energy and Industrial Strategy (BEIS). (2022b). *Twitter/X,* 7th April, p. 1.

Business, Environment and Industrial Strategy (BEIS). (2021a). *Twitter/X,* 24th November, p. 1.

Business, Environment and Industrial Strategy (BEIS). (2021b). *Twitter/X,* 29th October, p. 1.

Business Green. (2022). *Twitter/X,* 12th January, p. 1.

Carrington, D. (2022). *Twitter/X,* 21st January, p. 1.

Clark, N. (2021). 'ECO PLAN BOOST: Boris Johnson can create staggering 1.7 MILLION new jobs in his green revolution, experts claim', *The Sun,* 25th March, pp. 1–4.

Craig, J. (2021). 'Thousands of "red wall" jobs at risk without green investment, TUC warns PM', *Sky News,* 12th September, pp. 1–11, https://news.sky.com/story/thousands-of-red-wall-jobs-at-risk-without-green-investment-tuc-warns-pm-12405630

Crerar, P., & Horton, H. (2023). 'Threat to green agenda as Sunak mulls over "anti-Ulez" poll victory', *The Guardian,* Tuesday 25th July, p. 8.

Department for Transport (DfT). (2021). *Twitter/X,* 22nd November, p. 1.

Dobbins, T. (2022). 'Good work: policy and research on the quality of work in the UK', *House of Commons Library Research Briefing,* 6th June, pp. 1–42, https://commonslibrary.parliament.uk/research-briefings/cbp-9561/

Dorfman, P. (2022). *Twitter/X,* 3rd December, p. 1.

Draca, M., Li, J., Muller, S, Nguyen-Tien, V., Riom, C., & Valero, A. (2021). 'Are "green" jobs good jobs? How lessons from the experience to-date can inform labour market transitions of the future', Policy publication: Grantham Research Institute on Climate Change and the Environment, *London School of Economics and Political Science,* 28th October, p. 1, https://www.lse.ac.uk/

granthaminstitute/publication/are-green-jobs-good-jobs-how-lessons-from-the-experience-to-date-can-inform-labour-market-transitions-of-the-future/

Duell, M. (2020). 'Experts blow up Boris's hydrogen pipe dream: PM's plan to replace gas boilers by 2023 is "impossible" because only TWO hydrogen prototypes exist, a fifth of gas network will need to be relaid and EVERY engineer retrained', *The Mail Online,* 19th November, pp. 1–16.

Dundon, T. (2022). *Twitter/X,* 30th June, p. 1.

Edgerton, D. (2022). 'The woes of start-up Britishvolt should shock the UK out of its Brexit self-delusion', *The Guardian,* Friday 11th November, pp. 1–3.

Evans-Pritchard, A. (2023a). 'Britain should be wary of joining Europe's Gigafactory stampede', *The Daily Telegraph,* Friday 19th May, p. 24.

Evans-Pritchard, A. (2023b). 'Tata's Gigafactory is Britain's entry into the global electric car league', *The Daily Telegraph,* Friday 21st July, p. 23.

Faragher, J. (2021). 'Green jobs plan needs more clarity, say MPs', *Personnel Today,* 25th October, pp. 1–2.

Goodman, P. (2023). 'Tory net-zero sceptics sense the wind of change', *The Times,* Wednesday 26th July, p. 21.

GOV.UK. (2021a). 'Press release: Prime Minister's Ten-Point Plan kickstarts green investment boom', Prime Minister's Office, Department for International Trade, Department for Business Energy & Industrial Strategy, *The Rt Hon Boris Johnson MP, The Rt Hon Anne-Marie Trevelyan MP and The Rt Hon Kwasi Kwarteng MP,* 15th October, pp. 1–6, https://www.gov.uk/government/news/prime-ministers-ten-point-plan-kickstarts-green-investment-boom

GOV.UK. (2021b). 'Press release: UK's path to net zero set out in landmark strategy', Department for Business, Energy & Industrial Strategy (BEIS), Prime Minister's Office, 10 Downing Street, Department for Environment, Food & Rural Affairs (DEFRA), *Department for Transport and the Rt Hon Greg Hands MP (2021),* 19th October, pp. 1–12, https://www.gov.uk/government/news/uks-path-to-net-zero-set-out-in-landmark-strategy

Greenpeace UK. (2022a). *Twitter/X,* 7th April, p. 1.

Greenpeace UK. (2022b). *Twitter/X,* 2nd December, p. 1.

Greenpeace UK. (2022c). *Twitter/X,* 5th December, p. 1.

Greenpeace UK. (2022d). *Twitter/X,* 7th December, p. 1.

Greenpeace UK. (2023). *Twitter/X,* 4th January, p. 1.

Guterres, A. (2021). *Twitter/X,* 10th November, p. 1.

Hardman, I. (2023). *Twitter/X,* 13th January, p. 1.

Harrabin, R. (2022a). *Twitter/X,* 5th December, p. 1.

Harrabin, R. (2022b). *Twitter/X,* 21st January, p. 1.
Harvey, F., & Elgot, J. (2021). 'UK's net zero plan falls short on ambition and funding say critics', *The Guardian,* Tuesday 19th October, pp. 1–4.
Hoare, C. (2021). 'Brexit Britain to create 1,340 jobs with "major" new £266 million green energy project', *Daily Express,* 9th August, pp. 1–8.
Holder, M. (2022). 'ArcelorMittal, BHP, and Mitsubishi trial carbon capture technologies for steelmaking', *BusinessGreen,* 28th October, pp. 1–5.
Holland, L. (2021). 'Fears offshore oil and gas workers may lose jobs in transition to green energy', Sky News, 23rd June, pp. 1–11.
Horton, H. (2023). 'Therese Coffey "complacent" in dealing with water companies, peers say', *The Guardian,* Tuesday 19th September, pp. 1–5.
Houchen, B. (2022). *Twitter/X,* 8th August, p. 1.
Hubbard, M. (2022). *Twitter/X,* 3rd February, p. 1.
Hudema, M. (2023). *Twitter/X,* 3rd January, p. 1.
Hughes, M. (2023). 'bp, SeAH and Tees Valley Lithium among job generators on Teeside', *The Northern Echo,* 2nd January, pp. 1–4.
Hunt, J. (2023). 'The Right of our party is talking the UK down and it risks becoming a self-fulfilling prophecy', *The Daily Telegraph,* Saturday 20th May, p. 4.
Ingham, J. (2021). 'Green revolution can turbocharge UK economy and create 250,000 jobs', *Daily Express,* 8th February, pp. 1–7.
Institute for Apprenticeships & Technical Education (IATE). (2023). 'Apprenticeships: Sustainability Framework Overview', *Institute for Apprenticeships,* 16th April, p. 1.
Jacobs, S. (2023). 'Britain isn't in "managed" decline. The country is about to fall off a cliff', *The Daily Telegraph,* Tuesday 8th August, p. 14.
Jefferson MFG. (2022a). *Twitter/X,* 19th December, p. 1.
Jeraj, S. (2022). 'Meet the apprentices upgrading the UL's infrastructure for net zero: Young workers at the National Grid', *The New Statesman,* 18th October, pp. 1–4.
Johnson, B. (2021). 'Press release: second wind for the Humber Teesside and UK energy industry', Department for Business, Energy & Industrial Strategy (BEIS), *Rt. Hon Kwasi Kwarteng MP and Rt. Hon Boris Johnson MP,* 10th March, p. 1–8, https://www.gov.uk/government/news/second-wind-for-the-humber-teeside-and-uk-energy-industry
Jolly, J. (2023). 'Gigafactory gives UK battery sector potential at last', *The Observer,* Sunday 23rd July, p. 51.

Jones, C. (2020). 'The corporate powerhouses behind PM's green dream', *The Times,* Wednesday 18th November, p. 1.
Kantor, A. (2021). 'Green Finance: The Thunberg Generation Demands Jobs That Mirror Green Values', *Bloomberg,* 9th December, pp. 1–2.
Kapadia, C. (2023). 'Sunak must be ambitious and heed Biden's investment blitz on energy', *The Daily Telegraph,* Tuesday 13th June, p. 18.
Keating, C. (2022). 'Reports: Britishvolt secures last-minute funding boost to avoid collapse', BusinessGreen, 1st November, pp. 1–2.
Kennedy, J. (2021). 'Gov green jobs revolution needs to speed up by 25 times to hit target', *The HR Director,* 27th October, pp. 1–5.
Kwarteng, K. (2022a). *Twitter/X,* 20th April 20th, p. 1.
Kwarteng, K. (2022b). *Twitter/X,* 27th July, p. 1.
Lansley, S. (2023). 'Broken Britain and the return of "private affluence and public squalor"', British Politics and Policy at LSE, *LSE Blogs,* 22nd September, pp. 1–7.
Littlecott, C. (2021). *Twitter/X,* 3rd November, p. 1.
London School of Economics and Political Science (LSE). (2021). 'Green economy: how the transition to net-zero could affect UK jobs across the country', Grantham Research Institute on Climate Change and the Environment, *London School of Economics and Political Science,* 5th March, pp. 1–4, https://www.lse.ac.uk/granthaminstitute/news/green-economy-how-the-transition-to-net-zero-could-affect-uk-jobs-across-the-country/
Maidment, J., & Penna, D. (2022). 'Russia's invasion of Ukraine reason "to act faster" on climate change, Rishi Sunak says', *The Telegraph,* 7th November, pp. 1–6.
Marlow, B. (2023). 'Electric car debacle shows net zero sums don't add up', *The Daily Telegraph,* Friday 1st September, p. 20.
May, E. (2021). 'More work needed to create green jobs, report says', *BBC News,* 29th November, pp. 1–3.
McCulloch, A. (2021). 'Green revolution "risks being tripped up by lack of green jobs"', *Personnel Today,* 21st October, pp. 1–3.
McLoughlin, B. (2021). 'Green Brexit Britain victory: Jaguar turns electric as UK jobs saved – Remainers red-faced', *Daily Express,* 16th February, pp. 1–3.
Monbiot, G. (2022a). *Twitter/X,* 26th October, p. 1.
Monbiot, G. (2022b). *Twitter/X,* 7th December, p. 1.
Mums For Lungs. (2023). *Twitter/X,* 4th January, p. 1.
Murray, J. (2022a). '"We are creating a Green Britain": Number of green jobs in the UK almost trebles', Business Green, 12th December, pp. 1–2.

Murray, J. (2022b). 'New LinkedIn data reveals UK green hiring surge', *Business Green,* 16th March, pp. 1–2.
Murray, J. (2022c). *Twitter/X,* 13th October, p. 1.
Murray, J. (2022d). *Twitter/X,* 25th October, p. 1.
Murray, J. (2023). 'Summer sales surge sees one battery electric car registered every 60 seconds', *Business Green,* 4th August, pp. 1–8.
Mustoe, H. (2023a). 'Battery failures that left car industry on brink', *The Daily Telegraph,* Thursday 18th May, p. 18.
Mustoe, H. (2023b). 'Tata closer to UK Gigafactory decision', *The Daily Telegraph,* Friday 19th May, p. 21.
Northern Powerhouse Partnership (NPP). (2022). *Twitter/X,* 25th January, p. 1.
Oliver, M. (2023). 'Britain's carbon capture plans at risk without EU deal, Sunak told', *The Daily Telegraph,* Monday 28th August, p. 17.
Ormsby, B. (2021). Green and clean sectors could fuel 15,000 jobs for next generation, *The Business Desk,* 22nd October, pp. 1–2.
Paul, J. (2022). 'Sunak handed get out of jail free card as Rolls-Royce pledges to decarbonise planes', *Daily Express,* 2nd November, pp. 1–9.
Possible. (2022). *Twitter/X,* 21st February, p. 1.
Promoting People. (2022). *Twitter/X,* 10th January, p. 1.
Rail Delivery Group (RDG). (2021). *Twitter/X,* 5th November, p. 1.
Ralston, J. (2021). *Twitter/X,* Saturday 4th December, p. 1.
Riley-Smith, B. (2023). 'Labour's net zero plan needs 3,000 turbines', *The Daily Telegraph,* Monday 17th July, p. 1, https://www.telegraph.co.uk/politics/2023/07/16/labour-build-3000-wind-turbines-meet-starmer-targets-zero/
Robinson, J., & Maidment, J. (2020). 'Pay as you drive to save the planet: Rishi Sunak con-siders plan to CHARGE motorists for every mile they drive to fill £40bn fuel tax hole left by the early switch to electric cars – as the government prepares to launch massive "reset" green plan', *Daily Mail Online,* 16th November, pp. 1–9.
Scottish Government. (2017). *Green Jobs Increase,* 6th April 2017, https://news.gov.scot/news/green-jobsincrease
Scottish Trades Union Congress (STUC). (2022). *Twitter/X,* 17th January, p. 1.
Sharma, A. (2022). *Twitter/X,* 3rd December, p. 1.
Sikka, P. (2022). 'There is no economic case for austerity. Here's why', *Left Foot Forward,* 4th November, pp. 1–3.

Singh, D. (2018). *Green Jobs Horizon Expands in Rapid Climate Changes Era*, LinkedIn, 1st June, pp. 1–4.
Skidmore, C. (2022). *Twitter/X,* 23rd February, p. 1.
Skidmore, C. (2023). *Twitter/X,* Sunday 24th September, p. 1.
Sofroniou, N., & Anderson, P. (2021). The green factor: Unpacking green job growth. *International Labour Review, 160*(1), 21–41.
The Economist. (2023). 'Hug pylons, not trees', *The Economist,* April 8th–14th, p. 7.
Thomas, N. (2022). 'Renewable energy: Support urged for oil and gas workers wanting green reskilling', *Financial Times,* 20th September, pp. 1–3, https://www.ft.com/content/974babc6-4223-4c06-a2a9-87b23765575a
Timpson, J. (2023). 'On productivity, No 10 must get its act together', *The Daily Telegraph,* Saturday 12th August, p. 22.
Trades Union Congress (TUC). (2022a). 'Government not doing enough to create new green jobs, TUC warns', *FE News,* 17th February, pp. 1–3.
Trades Union Congress (TUC). (2022b). *Twitter/X,* 18th February, p. 1.
Trevelyan, A. M. (2021). *Twitter/X,* 19th October, p. 1.
UK Research and Innovation (UKRI). (2023). 'Impact of transition to net-zero could affect UK unequally', *UKRI,* 15th April, pp. 1–3, https://www.ukri.org/news-and-events/responding-to-climate-change/understanding-our-changing-environment/impact-of-transition-to-net-zero-could-affect-uk-jobs-unequally/
Unsworth, S. (2020). 'Green jobs can boost recovery and deliver levelling up agenda', *London School of Economics and Political Science News,* Monday 19th October, pp. 1–4.
Valero, A., Li, J., Muller, S., Riom, C., Nguyen-Tien, V., & Draca, M. (2021). 'Policy brief: Are "green" jobs good jobs? Lessons from the UK and EU to inform labour market transitions', Grantham Research Institute on Climate Change and the Environment/Centre for Climate Change Economics and Policy, *London School of Economics and Political Science,* October, pp. 1–8, https://www.lse.ac.uk/granthaminstitute/publication/are-green-jobs-good-jobs-how-lessons-from-the-experience-to-date-can-inform-labour-market-transitions-of-the-future/
Vaughan, A. (2017). 'Electric car revolution: calculating the cost of green motoring', *The Guardian,* Saturday 8th July, pp. 1–3.
Vaughan, A. (2023). *Twitter/X,* 3rd January, p. 1.

Walker, P. (2022a). 'Sunak accused of 'economic illiteracy' over pledge to block onshore windfarms', *The Guardian,* Wednesday 20th July, pp. 1–2.
Walker, P. (2022b). 'No 10 set to allow new onshore wind projects in England in U-turn', *The Guardian,* Monday 28th November, pp. 1–2.
Warner, J. (2020). 'If Boris's green agenda is to fly, it must deliver UK jobs and growth', The Telegraph, 6th December, pp. 1–3.
White, H. (2023). 'Rishi Sunak's five promises duck big challenges for government', *The Institute for Government,* 5th January, pp. 1–3.

4

The Green Jobs Taskforce: Introduction, Proposals, Tensions and Critique

Abstract This chapter sees me introduce the Green Jobs Taskforce (GJT) and outline and detail the GJT Key Proposals. I then critique the GJT itself and provide details and commentary on key stakeholder reaction to it from among the business and trade union communities.

Keywords Green • Jobs • Taskforce • Proposals • Critique • Commentary • Stakeholders • Reaction

Publication of the UK conservative government-commissioned Green Jobs Taskforce (hereafter GJT) in July 2021, at first sight, heralded a new chapter in British employment relations, with government, trade unions and industry leaders potentially working together to deliver a new green economy in a progressive, inclusive way not seen since the late 1970s. However, as my critical deconstruction of the GJT findings in this chapter reveals, this new dawn for UK green job creation has proved elusive and arguably become a missed opportunity.

D. W. S. Renwick, *Work, Politics and the Green Industrial Revolution*,
https://doi.org/10.1007/978-3-031-53067-8_4

The GJT Introduction

The original GJT working group commissioned in late 2020 developed over 300 policy ideas, which were 'distilled into 15 recommendations split across the three themes' of the 'green jobs lifecycle', and initially tested at multiple workshops. The GJT explicitly distanced itself from the UK government, stating that their views and recommendations expressed 'are those of the GJT and do not represent government policy', and added that they met with ministers four times to update them on progress (Green Jobs Taskforce, 2021a, b, p. 12). This circumstance seems puzzling, as it was the British conservative government under Boris Johnson who established the GJT and set their terms of reference. Thus, was the UK conservative government distancing itself from the GJT from the outset, and if so, why?

The GJT's Ministerial Foreword states that they 'will consider' the Taskforce's evidence and recommendations, and:

> We will ensure that these programmes are directed to support more people to get the skills they need to move into green jobs and consider where we might need to go further or faster to fill skills gaps identified by the Taskforce ... [with] the formation of a cross-cutting delivery group to drive coherent action across the green skills agenda, [which] cannot be the responsibility of government alone. We want to see businesses step up and invest in training the green workforce, and so we urge them to reflect on the Taskforce's work and use it to inform how they can benefit from and contribute to the green industrial revolution. (Anne-Marie Trevelyan MP and Gillian Keegan MP in Green Jobs Taskforce, 2021a, b, pp. 3–4)

From the outset, the GJT reveals that it is a consultation, scoping type exercise and report. Their Executive Summary Introduction continues that 'research suggests that one in five jobs in the UK (approximately 6.3 million workers) will require skills which may experience demand growth (approximately 10% of UK jobs), or reduction (also approximately 10%), in the transition', and that 'to grasp the opportunities afforded by a green industrial revolution, the UK must develop a comprehensive and holistic view of the green jobs and skills challenge' (Green Jobs Taskforce,

2021a, b, pp. 5–6). But is such increased British green job growth feasible? And do UK conservative politicians have enough policy and economic bandwidth to re/upskill 6.3 million workers too? The GJT numbers above look like a huge challenge to deliver in practice, and equal job losses explain why no net gain arises in UK green job numbers too (referred to earlier herein).

On 'Theme 1: Driving investment in net zero to support good quality green jobs in the UK', the GJT argue that 'the government should publish a detailed Net Zero Strategy to set out the policy measures to be taken over the next five to ten years to create market demand, unlock the public and private investment needed and, open pathways to green skills, education and training which should ensure that green jobs are good quality. A societally inclusive, economy-wide effort will be needed, with coordination of activity, at all levels' (Green Jobs Taskforce, 2021a, b, p. 7). Yet this statement seems concerning, as what happens next if such market demand is not created, relevant investment is not unlocked, pathways are closed and coordination is lacking? Their GJT and Net Zero Strategy seem too reliant on many other developments working well together while making assumptions about what is possible and achievable.

The GJT move to 'Theme 2: Building pathways into good green careers', which states that 'all levels of government, the education sector, industry, and unions, must therefore work together to address these longer-term skills gaps to ensure we are building lifelong pathways into good green careers, and the priority for industry should be to build on existing good practice to ensure people from all backgrounds can access green jobs' (Green Jobs Taskforce, 2021a, b, p. 7). They continue that to 'expand the talent pool, and bring people with a wider range of experiences into the green economy, where the skills, knowledge, and behaviours required for people to move into green jobs should be embedded into the curricula of relevant subjects at all stages in the learning cycle, with action to scale up existing capacity', and that 'a green careers marketing campaign should be launched to provide comprehensive and dynamic green skills and careers advice' (Green Jobs Taskforce, 2021a, b, p. 7). Such pathways building on UK pluralistic approaches to employment relations may be welcomed by the relevant stakeholders involved and arguably make a marked change to the prior UK conservative governments politically unitarist and unilateralist approaches of the 1980s and 1990s. However, can so many stakeholders drawn from such different political identities unite? Where is the previous, existing good practice to build on? And who grants access to green job careers, delivers new green curricula and scales up existing learning provisions? This section of the GJT seems quite light on answering such practical questions.

The GJT then details 'Theme 3: A just transition for workers in the high carbon economy', which states that 'the UK must manage the transition to net zero inclusively so that those affected workers have a voice in shaping the transformation' (Green Jobs Taskforce, 2021a, b, p. 8). This theme seems progressive, as it is akin to the International Labour Organization (ILO)-type vision of decent work, to the number 8 United Nations Sustainable Development Goal (UN SDG), and so to generating UK green jobs. The GJT continue that 'from a skills perspective, this means that industry, government, and skills providers should together ensure that the adult skills system can meet the challenge of the transition to net zero. This includes being responsive to local demand and supporting workers in high-carbon sectors to take opportunities in the new economy' (Green Jobs Taskforce, 2021a, b, p. 8). This GJT element showcases a technocratic, training/skills viewpoint and some small government intervention, as 'it is businesses in the high carbon economy that should continue to take the lead in driving forward de-carbonization [and] government should support businesses where needed' (Green Jobs Taskforce, 2021a, b, p. 8). But such thinking may mean limits to UK government action and more pressure on British-based businesses to lead. What if such UK businesses cannot, or will not, lead? The limitations of the British government and GJT's thinking to deliver a just transition for impacted workers emerge here.

The GJT states some 'next steps' in a series of recommendations to be enacted including: 'Theme 1: Driving investment in net zero to support good quality green jobs in the UK', which argues for the British government working 'with industry, unions and other key stakeholders to actively set out, as part of their Net Zero Strategy, how they will ensure that the green jobs created by employers are good quality regardless of skills base', and to 'establish a UK-wide body with national representation to ensure momentum and coherence on workforce transition, including progress in delivery supported by local transition bodies to ensure effective place-based strategies for the transition' (Green Jobs Taskforce, 2021a, b, pp. 8–9). Here, while after Hill (1997, pp. 66–70) notions of 1970s political corporatism may emerge, less is said about who will set up and lead such bodies, what their terms of reference are and who they then report back to in the British government. Such details may matter if only to avoid government policy drift and inaction (cf. Hogwood & Gunn, 1984), which some conservative UK politicians already sense (e.g. Frost, 2023a, p. 17).

The GJT continues in 'Theme 2: Building pathways into good green careers' that the British 'government, employers and education providers' should promote 'the effective teaching of climate change and the knowledge and skills required for green jobs', to 'work with the skills and education sector to attract and retain talented teachers to teach subjects, which are important for green jobs', to 'work together to ensure green careers advice and pathways are a continuous offer for all', and that the UK government 'should map, review and enhance other training pathways to ensure they support a diverse, inclusive and net zero-aligned workforce across the UK' (Green Jobs Taskforce, 2021a, b, p. 9). Yet questions arise over the keywords used here, i.e. how does the British government enact such promotion, work with, work together, map, review and/or enhance? The GJT gives us few practical answers to such questions.

Lastly, in their Introduction section, the GJT add on 'Theme 3: A just transition for workers in the high carbon economy' that 'industry, government, and skills providers should ensure the adult skills system can meet the challenge of the transition to net zero'. This includes being responsive to local demand and supporting workers in high-carbon sectors to take opportunities in the new economy, alongside working with unions 'to tackle barriers to retraining and up-skilling so that no worker is left behind'. Here, 'employers and sector bodies should set out business and skills plans engaging unions and workers and boost private investment and de-carbonization of industry, the government should prioritize supporting high carbon sectors to transition and increase productivity and competitiveness' (Green Jobs Taskforce, 2021a, 2021b, pp. 9–10). Still, less is said about the exact mechanisms either in place or forthcoming, to ensure how relevant stakeholders respond to these requirements. As such, while this GJT Introduction section contains some promise of progress towards new, decent UK-based green jobs, it also has many key faults. I now examine and analyse the specific, key GJT proposals in detail.

The GJT Key Proposals[1]

The GJT group was asked to 'advise government, industry and the skills sector on how to realize the UK's ambitions for green jobs', namely: 'the skills needed to drive a green recovery and reach net zero greenhouse gas emissions by 2050; how the UK can ensure green jobs are good jobs, and open to all; and how workers in high carbon-sectors can be supported to transition to the new green economy' (Green Jobs Taskforce, 2021a, b, pp. 11–12). Unfortunately, while this socially inclusive remit seems welcome, it is limited by a lack of detailed funding to deliver such goals. The GJT then asks 'What is a green job?',[2] and adopts a broad approach which goes beyond simply looking at low carbon sectors (Green Jobs Taskforce, 2021a, b, pp. 15–16), and identifying green sectors, and makes big statements that 'achieving net zero by 2050 will require a system-wide transformation of the economy' and that 'most occupations, to varying extents, will become green'. It then more modestly argues that 'research shows that not all sectors will be impacted in the same way, or at the same time, in terms of skill demands' (Green Jobs Taskforce, 2021a, b, pp. 15–16). However, this GJT focus on skills may obscure the role of genuine employee voice to help secure a just transition for workers moving towards a greener British economy, and the GJT roadmap to

[1] 'The Taskforce was announced on 12 November 2020 and forms part of the UK Ten Point Plan for a Green Industrial Revolution. It was convened by ministers from BEIS and the Department for Education (DfE), and is made up of members from industry, trade unions and the skills sector' (Green Jobs Taskforce, 2021a, b, p. 12).

[2] 'In this report, the term a "green job" is used to signify employment in an activity that directly contributes to - or indirectly supports - the achievement of the UK's net zero emissions target and other environmental goals, such as nature restoration and mitigation against climate risks' (Green Jobs Taskforce, 2021a, b, p. 15).

achieving it seems difficult to visualize without huge government and private sector/business funding for it.

The GJT next discusses 'the impact of the transition on the demand for skills in the UK' and 'well-established green sectors which will experience significant growth', including offshore wind, which, according to relevant industry estimates, 'will see an uptick in the demand for workers and skills, as by 2026 the sector could employ around 70,000 workers directly and in the supply chain, with employment growing 'in construction and installation, and in operations and maintenance, with demand strongest for technicians and engineers' (Green Jobs Taskforce, 2021a, b, p. 19). The GJT add that this sector 'will require a broad range of skills' and many skill gaps 'could be covered by workers currently in the oil and gas sector, given the transferability of their skills to offshore wind' (Green Jobs Taskforce, 2021a, b, p. 19). Yet the GJT's emphasis on many expected new green jobs in offshore wind seems limited here, especially considering the new British PM Rishi Sunak's support for fossil fuels in 2022–2023. Nonetheless, the GJT's stress on the hope of worker skills transferability from working in the oil and gas to offshore wind sector seems like some welcome news for employees desiring it, however limited.

The GJT continues that electricity networks 'must be upgraded and expanded to produce a larger workforce – with skills in both smart and traditional networks engineering', as a National Grid report 'estimates a need to recruit for 400,000 energy jobs between now and 2050 to get to net zero - 260,000 will be new roles, while 140,000 will be replacing those who have left the workforce due to natural attrition' (Green Jobs Taskforce, 2021a, b, p. 20). Thus, good news emerges with up to 120,000 new electricity network jobs, but a query also arises, as it is unclear if this National Grid figure is an over or underestimate. A similar but different picture emerges regarding smart systems technologies (SSTs), as the GJT argue that 'by 2050, the domestic market for SSTs and flexibility solutions could be supporting around 10,000 jobs, while the export potential in 2050 could [support] 14,000 jobs' (Green Jobs Taskforce, 2021a, b, p. 20). The GJT adds that the Construction Industry Training Board (CITB) estimate that UK retrofit 'will require 12,000 workers be trained every year for about the next four years, before the need to ramp up annual recruitment by up to 30,000 workers between years 2025 and

2030. This implies an increased trained workforce of up to 230,000 by the end of the decade, and a resulting need to urgently prioritize new recruitment and retraining' (Green Jobs Taskforce, 2021a, b, p. 20). Questions arising here include whether such British government and industry figures are accurate, as prior green job growth estimates have been queried before (Renwick et al., 2008), and staff shortages are emerging, as 'Currently, there is a critical shortage of retrofit designers and coordinators', coupled with specific low carbon training being 'required across all roles including every part of the construction supply chain' too (CITB, 2021 in Green Jobs Taskforce, 2021a, b, pp. 20–21).

The GJT states 'green sectors that are predicted to grow ahead of the transition', including new low carbon technologies, hydrogen and CCUS, with industry 'expected to be the largest single consumer of hydrogen by 2050' (Green Jobs Taskforce, 2021a, b, p. 21). But important differences exist between brown, blue and green hydrogen, the ability of any hydrogen source to meet net zero goals is unknown, and a technical danger exists in applying hydrogen to industrial use (as it is more flammable than gas). Additionally, the CCUS industry is criticized as being 'non-existent, unreliable and expensive' (Thunberg et al., 2023, p. 4), and using unproven (Marlow, 2023a, b, p. 22), ineffective and unworkable technology. The GJT adds that on skills gaps, 'tight timeframes may make recruiting and training sufficient new talent challenging. These skill gaps could however be partly covered by the current energy workforce, such as those in oil and gas, which already have many of the key skills needed' (Green Jobs Taskforce, 2021a, b, p. 21). Nevertheless, as such sector technologies are only emerging, the UK government will perhaps need them to scale up quickly and become rapidly proven to realize this GJT vision. Further, the 'adaptation skills' in the GJT may be seen as 'the climate resilience market is rapidly growing, in sectors such as housing and construction, water, infrastructure, local government, and nature conservation' (Climate Change Committee, 2019; 2020 in Green Jobs Taskforce, 2021a, b, p. 22).

Moving forward, the GJT seem to have a solid case in arguing that the 'sectors experiencing significant transformation' include the automotive sector, as 'retraining, upskilling and new recruitment' are needed quickly in the 'transition to electric vehicles (EVs)', due to rapidly rising EV sales in Britain (detailed earlier), which already reached some 1 million EVs registered by October 2022, and are some 17% of new car registrations (The Times, 2023, p. 25), and that the UK 'could produce around 1.6 million EVs per year by 2040 and establish seven giga factories in national territory' which 'could create up to 78,000 new jobs (both direct and in the supply chain)' (Green Jobs Taskforce, 2021a, b, pp. 22–23). But the closure of the UK flagship EV site at Britishvolt (also detailed earlier), and the UK government's lacklustre handling of it, raises serious doubts if such grand EV job plans will be fully realized.

Additionally on skills, as production operators and equipment technicians constitute '75% of a Gigafactory's workforce' they need level 2–3 qualifications, while '25% need graduate level skills retraining and upskilling affects 50,000 EV related workers', and 'an additional 7,500-10,000 workers will be needed in battery cell manufacturing by 2030' too (Green Jobs Taskforce, 2021a, b, p. 23). However, such changes leading to many new EV-related jobs by 2030 may only yield a relatively marginal gain in reducing British air pollution (for nuanced air pollution information, see https://www.mumsforlungs.org).

The GJT then mentions skills in heating and cooling, as by 2050 'the heat network sector could create between 20,000 and 35,000 direct jobs' (Heat Network Industry Council in Green Jobs Taskforce, 2021a, b, p. 24). Here, '60,000 workers needed training up in upskilling since the announcement of the UK government's Ten Point Green Plan ambition of 600,000 heat pump installations per year by 2028' (i.e. to install four million heat pumps by 2028). The CITB estimates that this generates a need for between 7500 and 15,000 heat pump installers a year to be trained within the next seven years, resulting in around 60,000 workers required, meaning most current heating system installers 'will need to upskill, as the skills required for this are significantly different from those involved in installing conventional heating systems'[3] (Green Jobs Taskforce, 2021a, b, p. 24). The GJT continues that 'the growth of circular economy sectors such as repair, remanufacture, refill, and servitization, could create between 54,000 to 102,000 net jobs across all British

[3] Including 'knowledge in heat loss calculations, hydraulic balancing, flow temperature calculations and heating system sizing' (Green Jobs Taskforce, 2021a, b, p. 24).

regions by 2030, and increase jobs locally, especially through repairs', and that 'Masters levels skills are needed including engineers, material scientists and managers, with the Deposit Return Scheme estimated to create between 3,000 and 4,300 jobs' too (Green Jobs Taskforce, 2021a, b, pp. 24–25). Yet, the success or otherwise of such large numbers of UK heat pump installations is not guaranteed, due to criticism of them in some British newspapers in 2022–2023. Issues here include the practicality/inconvenience of fitting UK heat pumps, affordability/high installation costs, weak heat flow, finding reliable and certified installers, and issues of replacing boilers as well as radiators. But such criticisms may also themselves be misplaced, due to more supportive UK press outlets finding British citizens seemingly wanting to reduce gas boiler use (because of high costs), and to avoid Putin's Russia as a gas supply source (due to his war in Ukraine).

The GJT charts job losses in the UK transition to net zero, with sectors like oil and gas experiencing a likely reduction in demand, as 'between 2014 and 2017, the British oil and gas sector lost over 70,000 direct jobs as well as those in the supply chain', and '80,000 workers are likely to leave the sector between 2018 and 2035 due to natural attrition' (Green Jobs Taskforce, 2021a, b, p. 25). However, such figures need constant updating, with new UK PM Rishi Sunak backing some pro-fossil-fuel jobs too. Additionally, while the GJT recognizes that job losses have occurred as COVID-19 impacted global oil demand and prices, they estimate that 'over 90% of the UK's oil and gas workforce have medium to high skills transferability and are well positioned to work in other energy sectors'. Here, examples include moves 'from oil and gas to decommissioning and subsea network projects, and medium transferability to offshore wind, CCUS, and blue hydrogen', with 'around 100,000 jobs in the UK offshore energy sector in 2030 projected to be filled by workers transferring from oil and gas and by new entrants from outside the sector', where the North Sea Transition Deal 'could generate a net increase of 40,000 direct jobs, even after long-term decline in the oil and gas industry is included' (Green Jobs Taskforce, 2021a, b, p. 12). But while such British green job creation figures may need much closer inspection and interrogation, subsequent good news for workers involved lies in some limited job security, as the GJT also mentions 'reskilling, upskilling and inclusivity, as a need to reskill the existing workforce' occurs, via new skills in 'decommissioning, energy diversification, and further digitalization of the sector' (Green Jobs Taskforce, 2021a, b, p. 25).

The GJT moves on to mention the 'importance of cross-cutting skills to the transition, including retrofitting and whole house retrofitting', to 'work across multiple technologies and optimize them including solar panels, EV charge-points, heat pumps, batteries, and smart systems to control these interacting technologies' (Green Jobs Taskforce, 2021a, b, p. 26). Here, STEM (Science, Technology, Engineering and Mathematics) skills underpin jobs needed in moving towards a green recovery, as scientists are required 'to innovate the technologies [and do] vital research on climate adaptation', plus engineers 'to utilize systems thinking to approach the complex challenges of decarbonization, and in enhancing the resilience of infrastructure and buildings to climate change' (Green Jobs Taskforce, 2021a, b, p. 26). Still, do enough UK-based retrofitters, scientists and engineers exist to undertake all this capacity, and can they complete it all? The demands on them to do it, and do so successfully, seem huge.

The GJT then discusses the 'digital and data skills' required 'to enable smart grid infrastructure to deliver a reliable system, and home heating systems using efficient AI heating controllers, and digital twin technology to improve efficiencies', plus 'building of energy infrastructure, especially in new technologies involving hydrogen and CCUS' (Green Jobs Taskforce, 2021a, b, p. 26). Yet a practical question arising here is whether Britain has enough technicians and workers to do all these things. And as they sound very complex, is the UK set up well enough to deliver them all and to do them excellently too?

The GJT talks about how project managers are required 'across all industries to manage timescales and budgets', meaning that 'traditional career pathways for project managers will also need appropriate incentives'. Here, 'educators, trainers, and assessors are needed to mainstream green skills provision across both the public and private sector', including communicators to implement original change, chart 'green career pathways and job choices', and facilitate 'collective technical advancements'. Plus, 'leadership, management, and communication skills to drive cultural change' towards a green economy, as 'new technologies, behaviours and systems need embedding' (Green Jobs Taskforce, 2021a, b, p. 27). However, do existing role holders (in project management, education, communications/change management and leadership/management) want to undertake all the changes to their jobs that the GJT details above? And can they do them all? Do they have the capability, capacity, commitment and consistency required to make such adjustments work? The changes proposed here by the GJT seem to ask a lot from such existing stakeholders.

The GJT then switches focus to detail how 'the impact of the transition will be different across Britain because certain sectors will benefit from a concentration of opportunity in specific areas', and as 'the growth of CCUS and low carbon hydrogen will lead to increased employment opportunities within the UK's industrial clusters' and 'demand for roles' in EV manufacturing and supply chains 'will benefit the manufacturing hubs within the midlands and other hubs' (Green Jobs Taskforce, 2021a, b, p. 28). Here, 'the percentage of jobs impacted could vary from 19% in London to 23% in the East Midlands, and Yorkshire and the Humber and Northeast Scotland still significantly dependent on the oil and gas industry', meaning that 'if poorly managed, the transition could have serious impacts on the broader local economy' (Green Jobs Taskforce, 2021a, b, p. 28). Thus, some big risks may emerge for workers concerned if this transition process is not managed well. On the upside, the GJT also state that opportunities exist in 'nature-based job creation' which could aid recovery in UK communities experiencing high unemployment before COVID-19, and 'three types of environmental development—improving woodland, peatlands, and urban parks—could potentially create over 16,000 jobs' in such affected locales, especially 'constituencies in the Pennines close to peatland, and likewise, constituencies with potential for seagrass restoration have a higher proportion of people on furlough and poorer employment prospects' (Green Jobs Taskforce, 2021a, b, p. 28). Whether or not such constituency jobs emerge depends upon such environmental development occurring, which is an open question due to the British government needing to fully fund it, and because previous UK conservative actions regarding it have not always been delivered, as evidenced in reduced tree-planting under the May, Johnson and now Sunak governments (Gibbons, 2023a, b, p. 4).

The GJT arguably seem quite controversial here, as they appear to imply that some (or even the only?) jobs available for unemployed British citizens (above) include tree-planting, woodland protection, and peatland/seagrass restoration. While such new, green jobs seem a very welcome development in the context of existing UK unemployment, the GJT seems to revert to the Youth Training Scheme (YTS) logic and thinking favoured in the 1980s by some conservative British politicians. Then, TINA (there is no alternative) arguments dominated, and in doing so, showed less thought about any stigma arising for unemployed youths doing such work. This prior stigmatization could now be repeated, as it seems that such unemployed UK citizens cannot perhaps get any of the higher-level and higher-status, green jobs detailed earlier. Do the GJT, only have or see, such tree/woodland plus peatland/seagrass jobs as the only jobs available for unemployed Britons? Do the GJT not want more for such marginalized UK groups? And how does this particular GJT element match their earlier stated wish to create 'decent work' for all? Unfortunately, this section of the GJT seems to have few answers to such questions. Indeed, as the GJT continues, 'many of the roles required to support the journey to net zero such as retrofit and rail are in sectors which are not location specific and will be needed across the UK. This suggests that training provision for roles within these sectors needs to be accessible throughout several regions' (Green Jobs Taskforce, 2021a, b, p. 28), and the GJT perhaps thus beg the question whether the British government will make the scheme above really UK-wide or not. Further, the GJT add that 'these anticipated employment impacts will be exacerbated by the fact that skills mismatches are already persistent across Britain, as some areas have a skills surplus while others see a deficit' (Green Jobs Taskforce, 2021a, b, p. 30), future regional UK skills mismatches may also appear.

The GJT moves on to detail that 'Local Enterprise Partnerships (LEPs) exist in a low skills equilibrium' and that 'to expand, businesses need to have access to the appropriate local skills or the ability to attract these skills into the area, so labour mobility plays a key role' for plumbers and builders, who are 'particularly vulnerable as they rely on local labour supply and can find it difficult to plan for current and future skills needs, and most labour mobility in the UK currently occurs within regions rather than between regions' (Green Jobs Taskforce, 2021a, b, p. 30). Still, while the GJT seem to acknowledge that greater labour mobility is needed to fill labour market shortages (and diagnose this problem), they state a few practical steps to help deliver such mobility (and tackle this problem).

The GJT then talks about how the 'impact of the transition will be felt at different times', as 'employment in transitioning sectors could decline at a faster rate than new green sectors emerge, and/or in different locations, stopping workers from successfully transitioning, promptly', which may lead to a range of important skills being under-optimized or lost from our economy. For example, the British oil and gas sector may lose 'up to 30,000 jobs within 12-18 months, increasing local unemployment and the risk of capability loss in high-skilled areas', and a clear direction in government policy is needed, 'including further clarity of UK decarbonization pathways, investment timelines and location' (Green Jobs Taskforce, 2021a, b, p. 30). This GJT aspect seems reasonable, as it honestly describes impacts arising for workers concerned, but it leaves a big question of when such clear direction in UK government policy will arrive. The GJT continues that 'among the stakeholders we engaged with, there was a broad consensus that this gap can be met, but optimizing the skills currently available, and supporting workers to find quality work requires diverse timelines for green job creation and skills demands being matched with those for training and education provision, infrastructure and technology build and roll-out time', and workers transitioning between sectors plus local capacity issues arising (Green Jobs Taskforce, 2021a, b, p. 28). So, while timelines appear important again too, little detail emerges on when they will be resolved, i.e. who does what, when, where and how?

The GJT makes a series of 'recommendations to government, industry and the skills sector', across three key themes using relevant case study data, labelled: (1) driving investment in green jobs and skills in Britain, (2) building pathways into good green careers and (3) ensuring a just transition for workers in high carbon sectors. I now briefly detail and analyse these three themes.

Enveloped in 'Theme 1: driving investment in green jobs and skills', the GJT argue that 'businesses and investors will drive green job creation, with much investment needed coming from the private sector. However, the government must provide the right framework to underpin future policy, create clear market demand, and inspire the confidence needed to unlock investment from business' (Green Jobs Taskforce, 2021a, b, p. 32). However, it is not clear what happens next if this GJT financing plan does not occur on the scale they indicated. The GJT add that 'to unlock the huge scale of private investment needed, the government must publish a clear, sufficiently detailed, and long-term plan for the UK's transition, setting out key policy measures to be taken over the next five to ten years and the strategy should signal how and where market demand will be generated, and how the right skills and training can be put in place at the right time to match this' (Green Jobs Taskforce, 2021a, b, p. 32). However, to expect such a clear and detailed long-term plan may be over-optimistic, especially as the British policy environment now works under great turbulence because the UK had three Prime Ministers (PMs) within three months, and among them two PMs in either two months or seven weeks alone (depending on how this is counted, due to government inactivity around the mourning period of the late UK Queen Elizabeth II). Thus, policy strategizing, formulation and enactment in such an unstable British polity is far from guaranteed (for more context and discussion see the next chapter herein).

The GJT add that 'by setting out a national roadmap the UK government can enable, encourage, and expect sectors and businesses to publish their plans to invest in the green economy. Crucially, it will also give skills providers the confidence they need to design and offer courses to build the net-zero workforce' (Green Jobs Taskforce, 2021a, b, p. 32). Nonetheless, this point appears to misunderstand former PM Boris Johnson, as he used short-term timelines based on public opinion polls, media and focus group data in his policy-making and Cabinet (which itself rarely had a formal agenda), and his successor PM Liz Truss's very short-lived economic growth ideas crashed the Pound Sterling. Further, the new PM Rishi Sunak is cautious about adopting any utopian-style green plans, as he may have to change policy directions due to his small House of Commons majority of only 70, i.e. any 'pop up' rebellions against him may well have a big impact on his existing and future policy agenda (see Balls, 2023, p. 2; Dickson, 2022).

The more positive news comes from the GJT focus on research and development (R&D) and British government action 'through the £1 billion Net Zero Innovation Portfolio, and initiatives such as the Automotive Transition Fund (ATF)', where 'government should develop an R&D strategy aimed at increasing investment in critical innovative green technologies' (Green Jobs Taskforce, 2021a, b, p. 33). This GJT proposal seems more reasonable if the UK government have such money to spend, and whether conservative politicians see it as an incentive to stimulate private sector green investments. The GJT add that 'the design and delivery of government's strategy for net zero should include an ongoing assessment of the supply of and demand for skills, which should be kept under continued review' (Green Jobs Taskforce, 2021a, b, p. 33), yet it is not clear if the British government will do this too.

The GJT lastly detail a series of 'priority actions' (Green Jobs Taskforce, 2021a, b, pp. 74–76), many of which appear practical and feasible, i.e. the milestones for Recommendations 1, 2, 12 and 13. Nevertheless, their other recommendations suffer from many faults, namely: matching G7 levels of green stimulus investments (Recommendation 3); making assumptions about the UK government announcing additional funding for adult green skills and FE expertise hubs (Recommendation 11); and vagueness regarding increased decarbonization in their Comprehensive Spending Review (Recommendation 14)—because where is the money to fund them all coming from? The GJT also produce very tight timeframes and unclear plans on how to establish a framework to understand quality work in the green economy (Recommendation 4); unclear strategy on how local transition bodies are to be established by stakeholders (Recommendation 5); very tight timelines for schools, local government and employers to co-create modular programmes to maximize accessibility (Recommendation 7); arguably unrealistic ideas on expanding green apprenticeships to cover other training pathways (Recommendation 10); and very tight timescales on industry certification being standardized into a common framework, including passporting arrangements for offshore sectors (Recommendation 12). The GJT concluded, in their last Recommendation 15, that 'the government should set out how it will provide funding equivalent to the EU's Just Transition Fund to support communities to diversify and develop alternative industries' (Green Jobs

Taskforce, 2021a, b, p. 76). However, working under the (then)-Boris Johnson government instability, and the new and unstable Rishi Sunak administration, it is not clear when, and even if, such major investments will happen. I now offer a short critique of some key GJT elements which are not as well-surfaced above.

Critiquing the GJT

As the GJT's stance is that 'since education and training are devolved' matters, where their recommendations make direct reference to policy, 'they should be understood as being directed at the government for England only, and that similarly, recommendations directly linked to energy policy will not apply to Northern Ireland' too (Green Jobs Taskforce, 2021a, b, pp. 12–13). But without a common, Britain-wide approach to education, training and energy policy, how will the GJT deliver decent, green jobs for all UK stakeholders concerned? This seems like a missed opportunity to drive up standards in British green employment after Brexit.

On the 'next steps' ahead, the GJT states that the UK government 'has committed to ensuring that these recommendations will be considered in the development of Britain's first Net Zero Strategy'[4] (Green Jobs Taskforce, 2021a, b, p. 13), a commitment which is a start here. The GJT then says that it appreciates 'the economic impact of the transition to net zero will likely involve some job losses primarily in fossil fuel technologies', and notes that 'it will be vital to minimize hardships associated with these disruptions, such as by retraining workers, locating new clean energy facilities in heavily affected areas wherever possible, and providing regional aid' (Green Jobs Taskforce, 2021a, b, pp. 17–18). Still, the GJT lacks exact, comprehensive and holistic proposals on how to retrain such UK workers, where to relocate these energy facilities and when to deliver this regional aid too.

The GJT continues that 'the impact of the transition on the demand for skills in the UK will vary across the economy, as research by the

[4] On 30th March 2023, the UK's *Net Zero Strategy* became the *Carbon Budget Delivery Plan.*

European Centre for the Development of Vocational Training suggests that the greening of the British economy will influence skills demand in three main ways: (i) structural changes that will lead to increased demand for some occupations and decreases for others; (ii) new economic activity that will generate new occupations, with a subsequent need for new skills, qualifications and training frameworks; and, (iii) many existing occupations and sectors that will experience "greening" changes within their jobs, which should require adjustments to current training and qualification frameworks' (Green Jobs Taskforce, 2021b, p. 19). This last point seems to matter, as the reduced training, employment and life chances of workers left behind in UK mining communities in the 1980s—after the miner's strike—is an arguably heartbreaking, historical injustice which has never truly been redressed (for context see Benyon & Hudson, 2021; Callinicos & Simons, 1985; Crick, 1985; Paxman, 2021). Moreover, involving UK labour groups to help design and deliver points (i–iii) just mentioned above has never been a strength of British conservative politicians, as Boris Johnson's earlier green jobs plan lacks significant input from national trade unions and workers and as his approach there has been continued in the GJT. For example, the GJT (established in December 2020) had only two key trade union representatives out of 17 members (12%), who were from the nuclear-power Prospect Union and the Deputy General of the British Trades Union Congress (TUC). However, these two union representatives were outnumbered by 15 other representatives on the GJT from companies such as National Grid, Nissan, Tata Steel, BP and Barratt Developments (housing). All such 17 member-representatives reported back to two Conservative UK politicians as GJT chairs (namely Gillian Keegan MP and Anne-Marie Trevelyan MP), who then *fed back* an action plan for the British government to consider in Spring 2021 (GOV.UK, 2021a, pp. 1–4, keep my emphasis). Thus, any UK trade union plans to obtain a seat at the top table where decisions are made over British government environmental transition policies seems relatively limited here (cf. Kalleberg, 2018, p. 188 in Baldry & Hyman, 2022, p. 97).

Key Stakeholder Reaction

UK industry reaction after the GJT report was mixed, as a Confederation of British Industry (CBI) report raised it as an example of how Britain can 'win the green growth prize', but the CBI also seemed frustrated, saying that the UK 'can't just address the issues "around the edges" of changes needed', as 'we need to "get more strategic and more ambitious"' (Confederation of British Industry (CBI), 2023a, p. 2). Specifically, the CBI argues that:

> Critically, we have a business environment that is supportive of the transition, with more UK-headquartered firms signed up to the UN's Race to Zero campaign than any other country. However, other nations have recognized the prizes on offer and are rapidly ramping up their plans and spending to attract the green industries of the future. We have the direction and goals, and now need to move the delivery to secure investment, supply chains and green jobs. We need our strategy for 2030, prioritizing green industries where we can have a genuine strategic advantage. Otherwise, we risk throwing away the economic opportunity of the 21st century. The CBI estimates that by 2030, a non-exhaustive list of 27 green growth 'prizes' could increase in value to contribute £37-57 billions of annual UK GDP alone. (Confederation of British Industry (CBI), 2023b, p. 6)

British trade union reaction to the GJT came via a UK Trades Union Congress (TUC) blog, which argues that:

> The task force's recommendations show the way to high-quality green jobs. However, there have been concerns that the climate action we have seen to date has delivered too few jobs in the UK and that some of them are low quality. We can't assume that green jobs will automatically be safe and decently paid. The UK also needs to protect existing workers in industries that need to decarbonize. Prospect senior deputy general secretary Sue Ferns, who was a Taskforce member, added "It is now up to the government to show leadership, back these policies, and take a big step towards the cleaner and fairer future we all want to see". (Trades Union Congress (TUC), 2021, pp. 1–4)

Such developments and reactions above reveal the current conservative UK government's political Achilles Heel, namely a lack of clarity on how to put their very broad-brush green policy ideas into action without many union and worker-focused inputs. Instead, this conservative-led GJT seems reliant on both private-sector and private-equity funding and, as per Letza et al. (2004), is one where the British state—and people—may now be 'the prime stimulus and risk-taker in technological innovation, from which the private sector draws risk-free benefit' (in Baldry & Hyman, 2022, p. 99), e.g. in UK government subsidies for renewable energy, industry gigafactories and CCUS technologies. This circumstance produces some very hard-to-reconcile tensions for the British conservative government to balance, which include (among other things) using a strong state or big government (to set the direction of travel for government green jobs policy) alongside large private-sector and private-equity funding (to provide finance to deliver such a policy). Doing so exposes contradictions in UK conservative government thinking, as they collectively do not seem clear on which exact, particular political state/non-state, public/private funding model is best to use. I now expand on these more overtly political aspects in the next chapter, which uses conservative political ideas to explain and critique the GJT findings (above) overall.

References

GJT Introduction

Frost, D. (2023a). 'I fear for the future of Britain, a country in danger of just giving up', *The Daily Telegraph,* Saturday 2nd September, p. 17.

Gibbons, A. (2023a). 'Tories "extremely unlikely" to fulfil tree-planting pledge', *The Daily Telegraph,* Friday 11th August, p. 4.

GOV.UK. (2021a). 'Green Jobs Taskforce Terms of Reference', 1st December 2020, pp. 1–4, *GOV.UK,* published and accessed 28th July 2021, https://www.gov.uk/government/groups/green-jobs-taskforce#background

Green Jobs Taskforce. (2021a). 'Green Jobs Taskforce: Report to Government, Industry and the Skills Sector', *Department for Business, Energy & Industrial Strategy GOV.UK,* pp. 1–82, published and accessed 14th July 2021, https://www.gov.uk/government/publications/green-jobs-taskforce-report

Hill, M. (1997). *The policy process in the modern state* (pp. 1–254). Prentice Hall. ISBN 0-13-269226-0.

Hogwood, B. W., & Gunn, L. A. (1984). *Policy analysis for the real world* (1st ed., pp. 1–289). Oxford University Press.

Marlow, B. (2023a). 'Shapps is gaslighting us with promises on energy', *The Daily Telegraph,* Wednesday 3rd August, p. 22.

The Times. (2023). 'Leading Article: Braking Point: The 2030 target for ending the sale of petrol and diesel cars is becoming unfeasible', *The Times,* Wednesday 26th July, p. 25, https://www.thetimes.co.uk/article/the-times-view-on-vehicle-policy-car-crash-gxvfzg6gl

Thunberg, G., Burnelius, L., Ackerman, S., Jannok, S., Korhonen, I., Hirvasvuopio, J., et al. (2023). 'Burning forests for energy isn't "renewable" – now the EU must admit it, *The Guardian,* Monday 5th September, pp. 1–7, https://www.theguardian.com/world/commentisfree/2022/sep/05/burning-forests-energy-renewable-eu-wood-climate

GJT Key Proposals

Balls, K. (2023). 'Red Wall MPs go up against Sunak on legal migration', *The Spectator,* Monday 3rd July 2023, pp. 1–3. https://www.spectator.co.uk/article/red-wall-mps-go-up-against-sunak-on-legal-migration/

Committee on Climate Change (CCC) (2019) *Net zero - The UK's contribution to stopping global warming,* at: https://www.theccc.org.uk/publication/net-zero-the-uks-contribution-to-stopping-global-warming/

Construction Industry Training Board (CITB (2021) *Building skills for net zero,* at: https://www.citb.co.uk/aboutcitb/construction-industry-research-reports/search-our-construction-industry-research-reports/building-skills-for-netzero/

Dickson, A. (2022). 'Rishi Sunak has Tory rebel trouble all of his own', *Politico,* 1st December, https://www.politico.eu/article/rishi-sunak-rebellion-problem-tory-party-uk-onshore-wind/

Gibbons, A. (2023b). 'Tories "extremely unlikely" to fulfil tree-planting pledge', *The Daily Telegraph,* Friday 11th August, p. 4.

Green Jobs Taskforce. (2021b). 'Green Jobs Taskforce: report to Government, Industry and the Skills Sector', *Department for Business, Energy & Industrial Strategy GOV.UK*, pp. 1–82, published and accessed 14th July 2021, https://www.gov.uk/government/publications/green-jobs-taskforce-report

Marlow, B. (2023b). 'Shapps is gaslighting us with promises on energy', *The Daily Telegraph*, Wednesday 3rd August, p. 22.

Renwick, D. W. S., Redman, T., & Maguire, S. (2008). 'Green HRM: A review, process model and research agenda', Discussion (White) Paper Series, number 2008.01 (April 2008), Management School, *University of Sheffield*, pp. 1–46.

Critiquing the GJT

Baldry, C., & Hyman, J. (2022). *Sustainable work and the climate crisis: The link between labour and climate change* (pp. 1–277). Routledge.

Benyon, H., & Hudson, R. (2021). *The shadow of the mine: Coal and the end of industrial Britain* (pp. 1–402). Verso.

Callinicos, A., & Simons, M. (1985). *The great strike: The Miner's strike of 1984–5 and its lessons* (pp. 1–256). Socialist Worker.

Confederation of British Industry (CBI). (2023a). 'How the UK can win the green growth prizes', pp. 1–7, *CBI*, https://www.cbi.org.uk/articles/how-the-uk-can-win-the-green-growth-prizes/#:~:text=Our%20recommendations,devising%20tax%20and%20regulatory%20policy

Confederation of British Industry (CBI). (2023b). 'GOING FOR GREEN: The UK's net zero growth opportunity', *CBI*, pp. 1–69, https://www.cbi.org.uk/articles/how-the-uk-can-win-the-green-growth-prizes/#:~:text=Our%20recommendations,devising%20tax%20and%20regulatory%20policy

Crick, M. (1985). *Scargill and the miners* (pp. 1–160). Penguin Books. ISBN 0-14-052355-3.

Letza, S., Smallman, C., & Sun, X. (2004). Reframing privatisation: Deconstructing the myth of efficiency. *Policy Sciences, 37*, 159–183.

Paxman, J. (2021). *Black gold: The history of how coal made Britain* (pp. 1–392). William Collins.

Trades Union Congress (TUC). (2021). 'A greener economy can be positive for workers too', *TUC*, pp. 1–5, https://www.tuc.org.uk/blogs/greener-economy-can-be-positive-workers-too#:~:text=Expanding%20unemployment%20schemes%20like%20Kickstart,over%20the%20next%20three%20years

5

Critiquing the GJT Using Conservative Political Ideas

Abstract In this chapter, I introduce conservative political ideas as a theoretical lens by which to view and critique the Green Jobs Taskforce (GJT), drawing on Barbara Goodwin's work on conservative political ideas and ideology. I then critically discuss the fate of the UK conservative political leaders involved before the GJT, during the GJT and afterwards, including the new British Prime Minister Rishi Sunak.

Keywords Conservative • Political • Ideas • Theory • Lens • Critique • Ideology • Sunak

Introduction

Using the theoretical lens of conservative political ideas is a less familiar and novel development in HRM, as it focuses on issues of external government regulation and organizational context, and essentially puts politics 'back into HRM' (cf. Cooke, 2017; D'Cruz et al., 2021; Farndale et al., 2023; Rodriguez et al. 2017; Vincent et al., 2020). Such ideas also help us to understand the inherent tensions and contradictions of the key UK political actors themselves, namely conservative British politicians

D. W. S. Renwick, *Work, Politics and the Green Industrial Revolution*,
https://doi.org/10.1007/978-3-031-53067-8_5

and their use of ideology and the UK state and government too. The British government has made slow progress towards building a new green jobs economy; they seem hostile to dismantling fossil fuel energy sources, wish to re-assert the rights of individuals to bypass a potential green 'nanny state' and arguably did the groundwork in enabling the right-of-centre UK populist 'Net Zero Watch' (NZW) group[1] to emerge as a political counter-force against the GJT, and the only limited acceptance of the GJT by the new PM Rishi Sunak. I now elaborate on these points drawing on the work of the British political philosopher, Barbara Goodwin (2014).

Recent UK conservative leaders like Boris Johnson seem to follow Wilson's notion of the 'conservative character', including a belief in rules and a practical approach—which is seen in his original Ten Point Green Jobs Plan and the GJT. Nonetheless, Johnson also pursued a characteristically un-conservative fixed goal in adopting Theresa May's legal target of achieving net zero carbon by 2050, whereas such plans and targets are normally associated with socialism, communism and utopianism, meaning Johnson wanted to show his credentials as an incremental reformer (cf. Goodwin, 2014, pp. 161, 163, 176). Indeed, while after Plato some conservatives believe that 'change equals decay', Britain's GJT requires radical change to the UK economy, businesses and society, which makes it a rather unorthodox conservative policy—as it accepts risky man-made change—even though the outcomes from such change cannot be well-predicted and 'proven desirable in advance' (cf. Goodwin, 2014, p. 163). The current British government, therefore, draw on Edmund Burke's (1790) claim that a state needs some change to survive and conserve itself, and attempts to do this as they seek to modernize UK capitalism through workplace greening, which builds upon existing societal 'concerns and traditions' (of the climate crisis) and reinforces existing conservative tensions between 'damaging and acceptable evolutionary change'

[1] The NZW is a non-charitable arm of the Global Warming Policy Foundation (GWPF). I include few NZW or GWPF works herein, as the GWPF is under investigation by the UK Charity Commission. Here, the GWPF is alleged to have breached charity guidelines and not acted for the public benefit by accepting donations from pro-fossil fuel oil and gas companies (see Bychawski, 2022), https://www.opendemocracy.net/en/watchdog-urged-to-strip-global-warming-policy-foundation-of-charitable-status/.

(cf. Burke, 1790 in Goodwin, 2014, p. 166). Thus, the British government use some notions of 'big government' and state intervention to realize their GJT, and in doing so appear to reject tradition—which itself builds upon conservative notions of 'the wisdom of past generations' and accumulated experience. Instead, the UK government appear to overturn such tradition by submitting 'to the small and arrogant oligarchy of those who merely happen to be walking about' (Goodwin, 2014, pp. 166–167), namely the so-called contemporary green experts of Theresa May, Boris Johnson and now Rishi Sunak.

In making such moves, the British GJT implicitly accepts a faith in human goodness, i.e. that UK-based firms, workers and citizens will adopt the GJT recommendations, which is the exact opposite of conservative pessimism on human 'weakness, selfishness, and irrationality, and thus sees such conservative politicians disagreeing with Thomas Hobbes' liberal suggestion that people require more authoritarian-style government (cf. Goodwin, 2014, pp. 168–169). However, the faith of these UK political leaders in British humankind seems problematic, as few guarantees exist that individual UK organizations, managers, workers, trade unionists and citizens will accept a 'green jobs revolution', and what happens next if such citizens reject the trade-offs it requires, namely of reduced personal freedoms and high financial costs. Thus, the question of 'what is to be done' with anti-green stakeholders looms large, as they may appear somewhat like 'untouchable outcasts' (cf. Goodwin, 2014, p. 170), primarily because the British GJT contains few ideas on how to work with such discontents. Here, new UK social, green-based class differences and political cleavages may emerge between (a) pro-greens versus anti-green opponents of big eco-government, such as the climate change sceptics and deniers in the NZW group, and (b) the new pro-green elite seemingly forming in British business and government circles versus ordinary, less-privileged UK citizens. Such social divisions could split British-based workforces, create 'marginalized groups with little power' (cf. Wilkinson et al. 2021, p. 2), and potentially increase already-existing UK social stratifications.

Hence any pleas from the current conservative British government for UK workplace stakeholders to see the 'big picture' of the climate emergency and adopt the proposals and recommendations in their GJT may rely on an unrealistic view of British social unity. This is because such gains may not materialize if some workers are made redundant because their job is 'not green', and conversely other workers gain 'eco jobs', i.e. elite, highly educated environmental enthusiasts having access to the best green education and training. Such new social divisions might undermine notions from UK New Right thinkers inspired by Roger Scruton, who emphasized conventional conservative values of 'allegiance, tradition, and national identity' (cf. Goodwin, 2014, p. 177; Scruton, 2017). Indeed, a political consequence of the conservative in egalitarian view of human nature 'accepts the need for elite government and leadership' without mass participation (Goodwin, 2014, p. 171), and this elitist-style outcome appears in Britain's GJT because, as stated earlier, few labour, union and workplace stakeholders seem to have fed their ideas into it. Here, enduring UK conservative convictions seem to continue, of 'upholding the supreme executive arm of government against interference by others e.g., trades unions' (Goodwin, 2014, p. 171). Nevertheless, this new circumstance also uproots the thinking of conservatives like Joseph De Maistre and Alexis De Tocqueville, who criticized democracy 'for its oppressive populism' (Goodwin, 2014, p. 171) and is perhaps a new situation which sees the British government seemingly keen to be populist and to prove that they connect with UK societal concerns regarding the climate emergency.

As such, the respective and relevant Boris Johnson and Liz Truss and now Rishi Sunak conservative British governments could present their GJT as a unifying political force, i.e. as the green jobs policy road-map ahead, and a new political consensus on tackling climate change through creating new, UK-based green employment (cf. Goodwin, 2014, pp. 171–172). From their perspective, as they have now—allegedly and apparently—reconfigured the British economy towards 'green jobs', others do not need to do so (e.g., trade unions and workers), which may also imply from their conservative perspective that such 'other' UK labour voices do not need to be heard. By controversially closing off such (un) voiced, alternative, union-led green job plans, these conservative British politicians—notably Boris Johnson and Rishi Sunak (due to Liz Truss's only short premiership)—have perhaps achieved several conservative goals, of 'restraining the unruly, avoiding chaos and anarchy, and maintaining order and stable government', which link to Edmund Burke's conservative 'rejection of delegation' (cf. Goodwin, 2014, p. 172). The UK's new, conservative pro-green political leaders may thus see British businesses, workers and citizens as having 'little control', because such citizen freedoms appear subordinated to the duty of following the UK government's GJT, i.e. as their duty as citizens to British society, and to each other. That is to say, their GJT implies no need for 'green welfare rights' (e.g., legal-style guarantees of job transfers into new, green(er) jobs and retraining/upskilling from the polluter industries) because it already and supposedly promotes UK social cohesion and avoids 'damage limitation through strong legal controls' (cf. Goodwin, 2014, pp. 172–174).

Although Britain's conservative-led GJT appears to advocate maximum government regulation of UK working lives regarding green issues, it also seems anti-state, because while they detail that it aims to leverage private investment to stimulate new green jobs, it only provides limited state funding of approximately up to £8–12 billion per year to do so. Hence this conservative British government-commissioned GJT appears contradictory, as it obscures the exact role of the state in workplace greening. Doing so is problematic, as Teeter and Sandberg (2016) reveal that the starting role in greening processes is a tricky challenge to master, and because European observers detail many complexities in using the different industrial relations systems involved too (cf. Brandl, 2014). Here, the UK's GJT does not seem specifically designed to build worker co-operatives and new bottom-up, sustainable forms of green(er) work organization as the International Labour Organization (ILO) advocates (Poschen, 2015). This is because doing so may see British conservative politicians themselves being labelled as 'utopians' and pursuing a green dream, which is perhaps a forbidden fruit for them to desire, never mind to actually and actively seek out (cf. Goodwin, 2014, p. 174).

Thus, while conservative theorists like Michael Oakeshott, Friedrich von Hayek and Karl Popper stress 'a search for political truth through trial and error, accumulated wisdom and modest change', the UK government's GJT appears to impose a rationalist green scheme on British society seemingly only put together by them as politicians (cf. Goodwin, 2014, p. 174). In doing so, while the UK government (like these same three conservative theorists) prefers to stress it as a series of 'practical principles',[2] i.e. their 15 GJT 'Recommendations', such British conservative politicians also appear dogmatic about it in the guise of offering strong political leadership too (cf. Goodwin, 2014, p. 161). Nevertheless, in allocating 'a minor place to the individual' in their green job plans and GJT, the UK government now arguably leaves British workers, businesses and citizens relatively unsupported and perhaps less keen to buy into such plans (cf. Goodwin, 2014, pp. 175–176).

Essentially, a peculiarity of the UK conservative government's GJT is that it allocates a key role and some funding for the British state to potentially help generate several hundred thousand new green(er) jobs, alongside also relying on leveraging at least three times as much private sector funding and private equity money to do so. Their plan can be explained by interventionist policy closely associated with trade unions seeing such unions 'provided with an established partnership role in industrial policy' (Baldry and Hyman, 2022, p. 91), which the current conservative UK government seemingly wishes to avoid. The flip side of such British-style political interventionism is deregulation, which has very sadly produced lower UK health and safety standards with 'tragic environmental consequences', such as the 2017 London Grenfell Tower fire which killed 72 people and the 'dumping of 20,000 tonnes of waste each year into the North Sea' (Baldry and Hyman, 2022, p. 93). I now offer a short take on current green job developments under the new PM Rishi Sunak, which I also later question and critique in the next, final chapter.

UK Conservative Political Leaders

British conservative party support for green causes 'has generally been vocal at election time, but hesitant and half-hearted in power' (McKie, 2023, pp. 8–9). Examples include Margaret Thatcher's pro-green 1989 UN speech dwindling by 1992, David Cameron's pictures with huskies in the Arctic and his promise to be "the greenest government ever", but him then scrapping a £1 billion CCUS scheme, restricting the building

[2] Indeed, the UK government published a report on how to 'nudge' the British public to go green (see Londakova et al. 2021).

of new onshore wind farms, handing tax breaks to shale gas and him telling aides to "get rid of the green crap" in UK government energy policies, and Boris Johnson expanding British road building, airport approvals and new North Sea oil and gas licences (McKie, 2023, p. 9). Such conservative so-called greening saw UK conservative political leaders such as Theresa May, David Cameron and Boris Johnson all also paradoxically claiming—to varying degrees of success—how green they were as PM too.

The current era of Rishi Sunak as PM has seen him act in a non-green way too, in approving the first British coal mine in decades, 'nearly one year after the UK previously lobbied to consign coal to history and provide coal for steel-making rather than power generation' (Seabrook 2022, p. 2). Yet, Sunak was additionally set to do a U-turn on wind and solar planning rules, as he has approved one new, large solar farm near Chelmsford, Essex, in 2023 (Evans, 2023, p. 1) and has promised a consultation on relaxing planning restrictions for onshore wind farms, and the Environment Secretary revealed that the proposed blanket ban on solar projects on farmland was to be binned (Murray, 2022a, pp. 1–2).

However, in following the old cliché that if a week is a long time in politics, then a year may perhaps be fatal, by Summer 2023 Rishi Sunak bowed to pressure from right-wing UK conservative politicians and voters and backtracked to adopt more onshore wind farms in England, which drew the 'wrath' of some British conservatives and allegedly plunged Sunak's conservative party into a 'civil war' over their climate policy (see Helm, 2023; Murray, 2023 for details). Here, as more UK conservative party green policy changes emerge, Sunak looks perplexed, pained and wholly unsure of what role the British state should play in delivering them. Rishi Sunak has promised not to "unnecessarily" add costs to meet green targets (in Seddon, 2023, p. 1) and he would "keep making progress towards our net zero ambitions and to strengthen our energy security" (in Riley-Smith and Maidment, 2023, p. 1); these statements are exactly the kind of rhetorical non-commitments which end up politically supporting the UK pro-fossil fuel status quo.

Thus, it seems little surprise that one seasoned observer of UK conservative (Tory) politics describes 2022 as a year of 'Tory chaos', asks if 2023 could be 'even worse', and argues that Britain 'has become trapped in a doom loop of perpetual crisis under the Conservatives [with] manic mayhem at the apex of UK politics', and new PM Rishi Sunak's 'surreal route to the top [was one] where some of his supporters said boring is back' (Rawnsley, 2022, pp. 1, 3, 4).

Other people management experts state that Britain in 2023 is now in a terrible state, with 'consistently poor political leadership over 12 years of people wholly unsuited for public service' (Cooper, 2022, p. 1), and that the UK now has a wider polity post-Brexit 'indulging in post-imperial fantasies', rather than being ready to 'learn from those mid-sized economies that are both richer and more equal than it' (New Statesman, 2023, pp. 1–2). More specifically on the British conservative party green policy, commentators argue that 'a war against "eco-terrorism" became institutionalized across government agencies, and repressive tactics were honed against environmentalists' (Potter, 2022, p. 1). Such harsh tactics included the David Cameron-era 'decision to slash energy efficiency funding, delay green building standards and the British Green Deal fiasco' and Boris Johnson's repeated failure to deliver on his promise of a 'Rooseveltian green stimulus package', which all reduced carbon emission reduction efforts (Murray, 2022b p. 4).

In terms of future conservative policies, outgoing PM Liz Truss commissioned a Net Zero Review (NZR) led by Chris Skidmore MP over 2022–2023 to investigate how to achieve net zero by 2050[3] more affordably, increase green infrastructure and jobs, and tie in with popular UK societal concerns of tackling rising energy bills (Gatten, 2023, p. 4). However, it is perhaps difficult to imagine such recommendations being enacted under Sunak's new and arguably confused and confusing approach to British green government policy. Here, his new UK conservative green leadership must face up to their prior work and outcomes, which even in the conservative British Tory-supporting press is criticized, as: 'the government would like us to forget its record of soaring bills, lack of investment, greenwashing and precarious national security, but the public is being gaslighted' (Marlow, 2023, p. 22).

[3] The former UK Energy Minister Chris Skidmore signed net zero by 2050 into British law to become a legally binding target in June 2019 (Lofting, 2023, p. 15), while the ban of sale on combustion engine cars by 2030 was introduced in 2020 by Boris Johnson and the first Ulez for inner London was devised by Johnson too (Wright, 2023, p. 21).

Hence the UK needs a new green industrial strategy. But will British conservative leader Rishi Sunak or Labour leader Starmer provide it? Sunak has experience in high office and government first as Chancellor and now as PM, but less interest in net zero, while Starmer has much interest in net zero and expansive, rapid plans regarding it, but less experience in high office and government. Are either of them capable of leading Britain's response to the climate emergency and securing UK-based green jobs that doing so may produce? I now offer a discussion and some conclusions forward.

References

Baldry, C., & Hyman, J. (2022). *Sustainable work and the climate crisis: The link between labour and climate change* (pp. 1–277). Routledge.

Brandl, B., (2014). 'Industrial Relations Since The Economic Crisis', verbal paper presented to the Work, Organisation and Employment Relations Research Centre (WOERRC), Management School, *University of Sheffield,* UK, 19th November.

Burke, E. (1790). *Reflections on the revolution in France,* (edited with an introduction and notes by L.G. Mitchell, 1993). In *Oxford World's classics edition* (pp. 1–326). Oxford University Press.

Bychawski, A. (2022). 'Watchdog urged to strip anti-net-zero group of charitable status: MPs, peers, climate scientists and leading authors back letter sounding alarm over Global Warming Policy Foundation', *openDemocracy,* 23rd May, pp. 1–5, https://www.opendemocracy.net/en/watchdog-urged-to-strip-global-warming-policy-foundation-of-charitable-status/

Cooke, F. L. (2017). Provocation paper: Concepts, contexts, and mindsets: Putting human resource management research in perspectives. *Human Resource Management Journal, 28*(1), 1–13. https://doi.org/10.1111/1748-8583.12163

Cooper, C. (2022). *Twitter/X,* 21st December, p. 1.

D'Cruz, P., Delannon, N., McCarthy, L., Kourula, A., Moon, J., Spence, L. J., & Gond, J. P. (2021). Human relations special issue call for papers: Contesting social responsibilities of business: Experiences in context. *Human Relations,* 1–7. https://journals.sagepub.com/page/hum/call-for-papers

Evans, S. (2023). *Twitter/X,* 26th June, p. 1.

Farndale, E., Bonache, J., McDonnell, A., & Kwon, B. (2023). Positioning context front and center in international human resource management. *Human Resource Management Journal, 33*, 1–16.

Gatten, E. (2023). 'Sunak announces new oil and gas licences', *The Daily Telegraph,* Tuesday 1st August, p. 4.

Goodwin, B. (2014). *Using political ideas* (6th ed., pp. 1–480). John Wiley & Sons Ltd.

Helm, T. (2023). Sunak U-turn on wind farms in England draws wrath of green Tories, *The Observer,* Sunday 2nd July 2023, pp. 1–10, https://www.theguardian.com/environment/2023/jul/02/sunak-u-turn-on-wind-farms-in-england-draws-wrath-of-green-tories

Lofting, N. (2023). 'How net zero was foisted on the British public', *The Daily Telegraph,* Wednesday 26th July, p. 15.

Londakova, K., Park, T., Reynolds, J., & Wodak, S. (2021). *Net Zero: principles for successful behaviour change initiatives: Key principles from past government-led behaviour change and public engagement initiatives,* Behavioural Insights Team (BIT), BEIS Research Paper Number 2021/063 (October 2021), pp. 1–55, https://policycommons.net/artifacts/1850370/government-net-zero-behavioural-plan/2597278/

Marlow, B. (2023). 'Shapps is gaslighting us with promises on energy', *The Daily Telegraph,* Wednesday 3rd August, p. 22.

McKie, R. (2023). 'Tory PMs promised the Earth but failed to deliver', *The Observer,* Sunday 23rd July, p. 9.

Murray, J. (2022a). '"A very positive signal": Sunak set to U-turn on wind and solar planning rules', *BusinessGreen,* 7th December, pp. 1–2.

Murray, J. (2022b). 'UK coal mine approval: Is this the worst ever climate policy decision?', *BusinessGreen,* 8th December, pp. 1–8.

Murray, J. (2023). 'The Tory civil war over climate policy is heating up – and there is no peace deal in sight', *BusinessGreen,* 3rd July 2022, pp. 1–8, https://www.businessgreen.com/analysis/4119365/tory-civil-war-climate-policy-heating-peace-deal-sight

New Statesman. (2023). 'Leader: How to reverse British decline: Rather than indulging in post-imperial fantasies, Britain should learn from those mid-sized economies that are both richer and more equal than it', *The New Statesman,* 4th January, pp. 1–2.

Poschen, P. (2015). *Decent work, green jobs and the sustainable economy* (pp. 1–182). Greenleaf Publishing Limited/International Labour Organization (ILO).

Potter, W. (2022). *Twitter/X,* 15th December, p. 1.

Rawnsley, A. (2022). 'After a year when the only certainty was Tory chaos, could 2023 be even worse?', *The Guardian,* Sunday 18th December, pp. 1–7.
Riley-Smith, B., & Maidment, J. (2023) 'Ban on new petrol cars by 2030 in doubt as Sunak refuses to back policy', *The Daily Telegraph,* Tuesday 25th July, p. 1.
Rodriguez, J. K., Johnstone, S., & Procter, S. (2017). Regulation of work and employment: Advances, tensions and future directions in research in international and comparative HRM. *The International Journal of Human Resource Management, 28*(21), 2957–2982. https://doi.org/10.1080/09585192.2017.1416555
Scruton, R. (2017). *Conservatism* (pp. 1–169). Profile Books Ltd. ISBN 978-1781257524.
Seabrook, V. (2022). 'First coal mine in decades approved – a year after UK lobbied to "consign coal to history"', *Sky News,* Thursday 8th November, pp. 1–8.
Seddon, P. (2023). 'Rishi Sunak vows not to add "unnecessary" costs to meet green targets', *BBC News,* Monday 24th July, pp. 1–10, https://www.bbc.co.uk/news/uk-politics-66290603
Teeter, P., & Sandberg, J. (2016). Constraining or enabling green capability development? How policy uncertainty affects organizational responses to flexible environmental regulations. *British Journal of Management, 28*(4), 649–665.
Vincent, S., Bamber, G. J., Delbridge, R., Doellgast, V., Grady, J., & Grugulis, I. (2020). Special issue: Situating human resource management practices in their political and economic context. *Human Resource Management Journal, 30*(4), i–iv, 461-623 (November), https://onlinelibrary.wiley.com/toc/17488583/2020/30/4
Wilkinson, A., Knoll, M., Mowbray, P. K., & Dundon, T. (2021). New trajectories in worker voice: Integrating and applying contemporary challenges in the Organization of Work. *British Journal of Management, 32*(3), 693–707. https://onlinelibrary.wiley.com/doi/10.1111/1467-8551.12528
Wright, B. (2023). 'Sunak's claims to be on the side of motorists are pure exhaust fumes', *The Daily Telegraph,* Tuesday 1st August, p. 21.

6

Discussion and Conclusions

Abstract This last chapter sees me critically discuss the UK conservative political legacy regarding both green jobs and the Green Jobs Taskforce (GJT), and (un)intended consequences arising for stakeholders involved. I then examine new conservative British approaches, UK press and media views, and Sunak's own green future, including his actions in his late summer 2023 reshuffle. I add commentary on UK Labour's green plan and ask where are we now with UK green jobs, and where are we going? I partly answer this question with a critical discussion of Rishi Sunak's key net zero speech of 20 September 2023. I close this study and book with some study limitations, policy and practitioner implications and future research ideas.

Keywords UK • Conservatives • Politics • Legacy • Stakeholders • Media • Sunak • Labour

The UK Conservative Legacy

Part of the new tradition emerging over the last few decades regarding the green concern historically within the British conservative political party (hereafter UK conservatives) is one of policy changes, budget cuts and

D. W. S. Renwick, *Work, Politics and the Green Industrial Revolution*,
https://doi.org/10.1007/978-3-031-53067-8_6

missed opportunities. Their new and changed policies after Brexit threaten UK environmental protections, as one leading water provision study argues that the British government's 25-year environmental plan produced a 'zombification' of UK environment policy. Additionally, no serious governance system enforcing previous European nature policies exists, former EU Habitats and Birds Directives are 'most at risk' (Manchester Friends of the Earth, 2018, pp. 1–3), and some see the UK conservatives as 'polluting your air' because they 'tried to avoid tackling killer air pollution that is so toxic it is illegal' (Morning Star, 2017, p. 1). Further, the conservative UK government seem set to adopt some already-decided outcomes, as Minister Michael Gove suggested that wind farms or nuclear power stations 'are the only clean energy sources that the UK government will consider' (Rayner and Swinford, 2017, p. 2), which the new PM Rishi Sunak supported, in backing "more nuclear or offshore wind" (Riley-Smith and Maidment, 2023, p. 1). Moreover, the UK conservative party's financial legacy is one where the years under former Chancellor George Osborne (2010–2016) saw the 'largest budget cuts of any UK government' fall on the environment department, which were 'equal to 57% in real terms over the course of two Parliaments'. Then, the financial settlement supposedly left 'other areas such as wildlife conservation, air quality and water pollution, forced to manage cuts of up to 80% over the two Parliaments, and to more than 5,000 redundancies' too (Howard, 2015, pp. 1–2).

More recently, the UK conservatives have moved from Boris Johnson's Ten Point Green Jobs Plan and GJT to Liz Truss's pro-fracking interventions and now onto Rishi Sunak's so-called ambition[1] to invest up to £20 billion of UK public taxpayers' money[2] into untested and unproven carbon capture utilization and storage (CCUS). However, Sunak seems unable to organize urgent upgrades to the electricity grid, as Britain moves away from gas for heating and towards using electric vehicles (Gatten, 2023a p. 4). Apparently 'green' policy initiatives are missed

[1] This is an ambition, because 'this £20 billion figure does not appear in Chancellor Jeremy Hunt's Spring 2023 Budget formal costings, or to be spent before the next UK general election in 2024/5, meaning it may either be dropped if the conservatives lose this election or be funded through levies if they win' (Electrical Review, 2023, p. 1).

[2] Roughly £3.5 billion per year between 2030 and 2040 (Gatten and Gutteridge, 2023, p. 1).

opportunities to invest more public money into home insulation, solar energy and onshore wind, and the latest in a long line of UK conservative environmental 'policy mistakes' or 'policy disasters' (cf. Dunleavy, 1995) which can be traced back to PM Theresa May's budget cuts to key green government departments, and later PM David Cameron's "get rid of the green crap" era (Murray, 2021, p. 4). Additionally, the UK conservatives' £3 billion clean air strategy 'did not include a car scrappage scheme, or address pollution from construction, farming and gas boilers, and passed [action on them] to local authorities' (BBC News, 2017, p. 5), i.e. a devolution of such powers to local government authorities some of which are run by opposition UK labour politicians. Such political devolution by the UK conservatives on green policy is advocated by Chris Skidmore's 2023 Net Zero Review, which forms an arguable political checkmate because local UK Labour politicians now also have to manage this new green circumstance.

UK conservative leaders and politicians may point to some green policy gains, namely: banning new diesel and petrol vehicles from 2040, a £255 million fund to help councils tackle emissions (BBC News, 2017, p. 1), the UK offshore wind sector success story (which is amongst the very largest globally), their original Ten Point Green Plan under Boris Johnson, the GJT, a £3 billion clear air strategy and Rishi Sunak's most recent CCUS investment ambition. Additionally, their focus on green policy has produced a large and significant pro-green Conservative Environment Network (CEN group within the UK Parliament). Here, within Sunak's Cabinet, he may need the support of such CEN MPs to get his other policies through the House of Commons, as former PM Boris Johnson's resignation and subsequent letter critical of Sunak's green policies have promoted a split in the UK conservative party. This development saw a new National Conservatism (NatCon UK) conference in May 2023 at Westminster addressed by Sir Jacob Rees-Mogg (see Frost, 2023a, p. 19; Somerville, 2023, p. 5), and also a different new group emerge called Conservative Democratic Organisation (CDO) (Elgot, 2023, p. 2), who wish to reinstall what they call "the" party democracy, i.e. grass-roots based UK conservative party members. This internal party split may not be good for the British green jobs agenda, as for the UK conservatives 'there is no obvious ideological homogeneity on show'

(Daily Telegraph, 2023a, p. 15), and for many others it means more a fractured political debate, infighting and navel-gazing on environmental policy instead of the government being laser-focused on enacting it (BBC Newsnight, 2023a). Thus, this development conveys the fate of all the different politicians involved, including prior UK conservative leaders such as Boris Johnson and Liz Truss.

(Un)Intended Consequences Arising

A tragedy for British workers emerging from the UK conservative legacy above lies in employee voices not being seen or heard much in formulating the original British green jobs plan by Boris Johnson, i.e. a lack of opportunities for such UK citizens, plus the constricted role for, and marginalization of, worker's groups within the British GJT itself. Such labour exclusions even extend to the UK and Irish HR profession, the Chartered Institute of Personnel and Development (CIPD), not formally being on the GJT itself either! Ergo, Boris Johnson's green jobs plan and his GJT have the unintended consequence of looking like yet another employee voice 'failure' in UK HRM studies (cf. Wilkinson et al. 2021). Indeed, if British workers are not trusted to build and drive forward a new green(er) UK economy, then who will? For all its fanfare, the GJT fails to answer this key question and is thus a missed opportunity to gain progress in British workforce greening. The GJT report also suffers from significant implementation and delivery issues, and a lack of serious, sustained levels of new long-term UK government funding to support new British green job creation. Thus, the findings in this research monograph, therefore, raise important questions on which stakeholders will now 'buy into' the GJT, and if not so many do so, whether it is instead seen as a stereotypical, classic political ruse, i.e. who will champion, enact and fully deliver the GJT recommendations after the GJT's original organizer, former UK PM Boris Johnson, left the political scene?

As such, the Boris Johnson and Rishi Sunak governments have not seemingly got 'green jobs done', because of the limitations of British state interventionism, which was and is itself facilitated and advanced by a conservative anti-state ideology and the arguably chaotic personality and leadership of both Johnson and Sunak too. Ergo, failures in the UK green jobs plan and the GJT arise, and with more successful counterparts being enacted elsewhere in the US and the EU, the British GJT seems like a key, missed political opportunity to generate new UK-based green employment with deleterious consequences for UK worker voice(s). Instead, the failure of the GJT now opens the door to more hawkish, extreme British right-wing conservative climate-sceptic voices and their UK media/press supporters (detailed earlier) to potentially fill the political vacuum emerging. Such voices perhaps include Steve Baker's Net Zero Watch (NZW) group, the Global Warming Policy Foundation (GWPF) and Conservative Way Forward (CWF) (cf. Pickard-Whitehead, 2023, p. 7), as all three groups seek to engage in anti-green rhetoric and political spin to support a classic conservative, highly political and ideological 'business-as-usual' no change type scenario, which stimulates UK-based fossil fuel industry players and only enacts decent, green jobs at the margins. In doing so, all three political groups seemingly do little to drastically alter the status quo of British industry and move it away from historic stagnation and decline and into a brighter, greener future. Any such class antagonisms arising from the pro-green/anti-climate tensions seen above may 'erupt in a disruptive way' (cf. Goodwin, 2014, pp. 170–171), which Steve Baker MP argues, as he states that due to their high costs and business and public affordability issues, the UK government's green job proposals 'could provoke a political crisis greater than those around the Poll Tax' of the late 1980s and even spark riots in Britain (Baker in Bet, 2021, p. 6). But Baker's NZW group seems like a climate sceptic one, and as a potentially future UK conservative party leader, he might arguably be seen to be encouraging more authoritarian-style governance.

However, a clear majority of Britons also voted for a net zero environment policy in the last 2019 UK general election, British uptake of electric vehicles (EVs) is at an all-time high, and most UK citizens 'say they want the environment prioritized in the economic recovery', and climate change has topped the list of national UK concerns for a third time (ipsos.com, 2020, p. 1). Further, a majority (54%) of Britons think jobs and investment in renewables are 'just as important' as securing low-cost power, and 11% want to prioritize the industrial/job benefits of the sector' (Bennett, 2022, p. 1). Indeed, an Office for National Statistics survey finds climate change and the environment ranked third in 2023 after the cost of living and the health service (Toynbee, 2023, p. 2), and some commentators add that onshore wind is: 'a) massively popular with Tory voters, b) a Great British invention, c) cheap d) and low carbon' (Bell, 2022, p. 1). Yet recent polls in the summer of 2023 also show 51% of Britons 'would like to do more to help' the climate but 'cannot afford to' (Gibbons, 2023, p. 6). Hence UK politics and public opinion matter, especially regarding the role of the British state, as the US, the EU and China have all seen much state intervention in green policy and are all leading the way on green job growth, while less interventionist and more cautious regimes like the UK government are struggling to keep up (Leake, 2023, p. 18). Here, the left-leaning British Institute for Public Policy and Research (IPPR) think-tank argues that the UK needs to spend £25 billion until 2032 on green manufacturing industries to achieve a comparable funding level to US and EU green industry investments (at £137 billion and £134 billion each respectively), as not investing in them means importing such green goods (Leake, 2023, p. 18).

New Conservative British Approaches

As a postscript herein, it seems pertinent to say something about the new UK PM Rishi Sunak's own green credentials and policy programme, and new Labour policies under leader Sir Kier Starmer too. But this is a very rough sketch below at the time of writing (late Summer 2023), as we all need to know a whole lot more about them to comment more and to do so both fairly and wisely. For now, my initial thinking is that the new Rishi Sunak-led British conservative government appear to inspire little confidence either of a green policy refresh or of a new start regarding the UK green jobs policy or of his leadership and regime helping to generate many more new, decent British-based green jobs. Why? This is because he: went ahead with opening a new coal mine in Whitehaven, Cumbria, UK; facilitated 100-odd new oil and gas drilling licences in the North Sea; earmarked up to £20 billion into carbon capture use and storage (CCUS) technology; rarely mentioned the climate crisis as Chancellor, and rarely mentions it much now as PM; has not stopped development of the Sizewell C nuclear power station; has opened up permits for more short-haul internal flights; took a plane, and not a train, to travel 80 miles from London to Southampton; and had to be embarrassed into attending the United Nations COP conference (Renwick, 2022, 2023). Moreover, Sunak is reported to have said that he could abandon green policies that cost British consumers—as he "doesn't unnecessarily [want to] give people more hassle and more costs in their lives [because] that's not what I'm interested in and prepared to do" (in Crerar, 2023, p. 2), he cancelled his government's pledge to enact a law banning UK coal power generation from October 2024 (Oliver, 2023a, p. 14), his senior aides say that Sunak's approach to net zero by 2050 is that "he's not opposed to it – he's just not interested" in it (Smyth, 2023a, p. 1), and he approved the controversial Rosebank oil and gas field in September 2023. Further, an internal government audit found green measures being 'allowed to run off course' under Sunak's tenure, including reducing packaging, plastic bottle returns, recycling initiatives, tree planting and woodland creation (Savage, 2023, p. 1).

Here, it is not that Sunak is deviating from official UK conservative party policy (i.e. achieving net zero by 2050), as he often talks about doing so. Rather, it is more that he is implementing it differently, because he seems decidedly non-green or neutral in his character and his changing, personal behaviours, and perhaps does not know what to say or do regarding the ecological concern. This impression was further enhanced in mid-2023 when Sunak had little reply to the stinging riposte by the British climate watchdog (CCC) that his government's progress on net zero is "worryingly slow, and markedly less" than in 2022 (in Thomas-Peter, 2023, p. 1), the resignation of long-standing UK climate Tsar Minister Zac Goldsmith, whose excoriating criticism is that Sunak is "simply uninterested and practices apathy" regarding the environment (in Scott, 2023a, p. 1), and comments by resigning government minister Nadine Dorries MP that "you simply don't care about the environment" (in Sky News, 2023a, p. 5), which for Zac Goldsmith imply that Britain had now "stepped off the world stage" (in Tominey, 2023a, p. 21). Additionally, other UK conservative politicians like former COP 26 President Alok Sharma MP argue that Britain needs £327 billion of investment to reach net zero by 2050 but has only committed £22.5

billion, and Chris Skidmore adds that "to follow a 'not zero' path would lead not only to further environmental crisis but a future economic crisis too" (in Helm and Harvey, 2023, p. 15).

Indeed, Sunak appears 'defensive and petulant when criticized' on television (Shipman, 2023, p. 7), and he seems much more comfortable talking about business and finance policy issues, which is predictable considering his background as a hedge fund manager and his Stanford MBA education, and his climate apathy may be influenced by or come from Donald Trump's climate scepticism/denial (detailed in Chap. 1) seeping into UK politics via his fellow politicians in the conservative NZW group and other UKIP politicians too (Logan, 2017, p. 1). Certainly, Sunak's announcement of billions for CCUS was made when British Parliamentarians were in recess (not attending), so no debate on it was possible (BBC Newsnight, 2023b); in late summer 2023, Sunak said that he is "on the motorist's side" (Malnick, 2023a, p. 1) and that his use of a private jet to make internal UK flights "is the most efficient use of my time" (Grant, 2023, p. 15). But such actions appear risky for Sunak, as some see that his government seems to be politicizing and vilifying net zero (Atkins, 2023, p. 1), and others state that Sunak is not acting in the national interest (Murray, 2023a, 2023b). Indeed, his latest oil and gas plus pro-CCUS drive appears tied together, which is problematic, because concerns arise that CCUS 'is viewed by oil giants as a "free lunch" to excuse further drilling rather than decarbonizing' (Wheeler et al. 2023, p. 2). Such outcomes are perhaps expected, as *The Guardian* newspaper links Sunak's family firm to oil giant BP[3] (Walker, 2023a, p. 1).

Sunak is now engaging in politicking and electioneering ahead of the next UK general election in 2024, as he characterizes his political opponents in Labour as 'Just Stop Oil' protestors (Smyth, 2023b, p. 10), perhaps blocking progress, supporting strikes, being dreamers and utopians, obsessed with unrealistic climate targets and having fixed ideas, while he alternatively portrays his own UK conservative party as motorist-friendly, flexible, making climate progress, being pro-trade, realistic, reducing costs and therefore reliable and trustworthy. So, Sunak is trying to put

[3] While Sunak's former Energy Minister Theresa Villiers MP also had shares in Shell (Sky News, 2023b, p. 1).

clear dividing lines between himself and UK Labour politicians, working to shore up his position with his supporters (about being pro-motorist and pro-oil and gas), looking to gain new votes in Scottish oil constituencies, and wanting to win more votes overall through his concern on reducing British voter energy bills, and relating his green policy to 'pragmatism'—which might resonate with UK public notions of 'common sense' policymaking (see *The Daily Telegraph,* 2023a, p. 15).

However, costs also arise for Sunak in pursuing green policy as an electoral 'wedge issue' and arguable culture war (Mahmood, 2023a, pp. 1–2), as the British think-tank Onward estimates any rollback on green targets could cost Sunak's conservatives up to 1.3 million votes at the next election, or a quarter of their current voters, and put off 'millions of wavering supporters' too (Payne, 2023, p. 19). Additionally, a 2023 Opinium poll revealed most UK conservative (Tory) voters planning to switch to Labour feel Sunak 'has not done enough on climate change' (Vaughan, 2023a, p. 8), and UK Tory chiefs worry about losing up to around 1.7 million rural votes from households who cannot afford to replace their oil heating systems (Somerville, 2023, p. 4). Here, are Sunak's conservatives using such similar 'votes won versus votes lost' figures as a smokescreen to push through their own more 'realistic' and pro-fossil fuel policies? I say this as the most recent climate discussion in the serious UK broadsheet conservative-supporting political press and media circles has now seemingly descended to new a low point, which has six parts detailed below.

UK Press and Media Views

Firstly, a steady drip-feed of arguably negative noise and news from UK conservative politicians exists (The Daily Telegraph, 2023b), such as Minister Suella Braverman MP in *The Daily Telegraph* labelling British Labour leader Starmer as being "in bed" with pro-climate group donors such as Just Stop Oil because he gained £1.5 million from one of their backers (see Gutteridge, 2023, p. 4), yet Starmer also said such groups should "just stop", and accused them of "arrogance" in their pro-climate stunts (Penna, 2023a, p. 4). Here, Sunak's former UK Energy Secretary Shapps added that Labour has "an ideological vendetta against British energy independence [which] risks hundreds of thousands of jobs" after industry sources claimed that Labour announced it would stop all new drilling licences in the North Sea if it wins the next election (in Oliver, 2023b, p. 4), which Starmer then rejected, as it may mean fewer jobs and votes in areas he needs to win (Daley, 2023a, p. 20). Starmer is also characterized as being 'a cheerleader for a tiny, tiresome pressure group' (The Daily Telegraph, 2023c, p. 15), and an apparent 'revolt' is happening from London Labour MPs over the expanded Ultra-Low Emission Zone (ULEZ) scheme despite Starmer saying it was a "difficult decision that had to be made" (Diver et al. 2023, p. 6). Here, conservative Shapps states that Labour's energy policy "represents a clear and present danger to national security" and that Labour has "anarchist allies", an "insane plan" (Shapps, 2023a, p. 5) and a "green underbelly" (in The Sunday Times, 2023, p. 24). Some conservative media commentators add that Labour has 'reformist rhetoric', whereas 'there's nothing there', and that Labour's Rachael Reeves Green Prosperity Plan 'was effectively shelved. Quite an embarrassment' (Nelson, 2023a, p. 14). Further, others state that Labour's net zero plans need '3,000 onshore turbines' to work too (Riley-Smith, 2023, p. 1).

Secondly, offshore British wind farm development executives describe former Energy Secretary Shapps 'as a "remote" figure who is reluctant to engage with company bosses' (Chan and Oliver, 2023, p. 27), and someone not always keen to listen to threats to such large wind projects like increasing costs, obscure planning rules and reduced subsidies, while the latest round of auctions for British renewable energy licences has 'hit the rocks', meaning 'slowly, but surely, the government is watering down its environmental agenda' (Warner, 2023a, p. 21). Instead, UK 'ministers remain bullish about the wind industry's prospects, with Shapps dismissing new offshore wind competition from the US' because "we are decades ahead" (in Oliver and Chan, 2023, p. 31).

Thirdly, right-of-centre British commentators, politicians and media outlets are characterizing (and stereotyping?) green enthusiasts as a 'snobby climate-change-deranged middle-class' and a 'Tesla army' (Long, 2023, p. 23), 'metropolitan elite green zealots' (McCartney, 2023, p. 19), the 'laptop elites of Davos' (Orlowski, 2023, p. 18), and groups where 'environmental extremists' practise 'eco-fascism' and a minority of immoderates exists (Tominey, 2023a, p. 21). They add that net zero is 'terrible for the economy', and 'there is a 'backlash' against it (Marlow, 2023a, p. 34; 2023b, p. 20), as a 'no growth green group' emerges (Riley-Smith and Martin, 2023, p. 1). They describe 'unpopular green net zero policies' (Frost, 2023b, p. 5) and 'net-zero dogma' (Moore, 2023, p. 18). UK Minister Michael Gove asks not to treat "the cause of the environment as a religious crusade" (in Malnick, 2023b, p. 1), and greens are seen to engage in 'apocalyptic alarmism' (The Times, 2023b, p. 23). Sir Jacob Rees-Mogg MP comments about 'unpopular, expensive green policies' (in Penna, 2023b, p. 6), the net zero agenda is 'filled with greenwash solutions that achieve nothing' (MacKinlay, 2023, p. 6), 'middle-class zealots and extremist environmentalists' exist (Tominey, 2023b, p. 19), who practise 'climate hysteria' (Daley, 2023b, p. 20), and that 'we will soon face energy outages, electric cars that can't be charged and heat pumps that don't work' (Heath, 2023a, p. 16). Moreover, other evidence is that lobbyists for gas boilers are attempting to delay UK government-led household heat pump installations, and instead promote hydrogen to extend the life of the gas industry regardless of hydrogen's unsuitability (i.e. hydrogen may be expensive, inefficient, leaky, dangerous and high-carbon) (Harvey and Cooke, 2023, p. 2).

Fourthly, the developments above sit alongside countless anti-net zero editorials and other pieces in the leading UK conservative political newspaper, *The Daily Telegraph,* including UK Labour being 'wedded to disastrous ideas on environmentalism' (2023d, p. 19) 'using a cloud of extravagant rhetoric' (2023e, p. 15), and there being 'an opportunity here for the government to create a new dividing line in British politics and be on the side of the ordinary voter by prioritizing cheap energy and reliable supplies' (2023f, p. 19). They comment on 'the net-zero poll tax' of green levies on citizens, being a question of whether 'our politicians are intent on making people poorer in the name of net zero?' (2023g, p. 17), criticize 'all these pointlessly punitive measures' (2023h, p. 17), and request to 'change the ludicrous net zero timetable' (2023i, p. 15). They add criticism of 'anti-car zealots' (2023j, p. 19), 'arrogant, unthinking greenery' (2023k, p. 17), that 'Rishi Sunak is finally standing up to the green Blob' (2023l, p. 17), and 'how the political Left captured the Royal Society for the Protection of Birds (RSPB)' (Heffer, 2023, p. 1). Additionally:

> Making net-zero legally binding by 2050 must count among PM Theresa May's very worst acts[4] … where is the cost-benefit analysis of net zero?… the conservatives should consider offering a referendum on net zero. (The Daily Telegraph, 2023m, p. 17)

To counter, former ministers like Alok Sharma MP urge the UK government to keep the political consensus on achieving net zero, and British business leaders, scientists and politicians warn that diluting climate policies would be very unpopular with voters, undermine global efforts towards net zero and hurt Britain's green credentials too (Savage et al. 2023a, b, p. 1).

Fifthly, other net zero criticisms include ones that it 'risks blackouts' (Gatten, 2023b, p. 8), contains 'irrational environmental schemes and eco-authoritarianism' (Johnston, 2023, p. 14), and that 'Scotland is now the testing ground for net-zero madness, and green policies that prioritize ideology over living standards' (Harris, 2023, p. 23). Electric vehicles (EVs) are now seen to cause a 'pothole crisis', due to EVs weighing more and causing more wear applied to roads through asphalt cracking and expanding (Simpson and Clark, 2023, p. 1), and a generation of children are being fed a never-ending "dark green" gloom narrative (Nelson, 2023b, p. 16).

Sixthly, some arguably more climate curious and/or climate sceptic commentators like former *Spectator* editor Fraser Nelson argue that 'the planet's future is brighter than many think', and 'good news' exists, such as fusion research in Oxfordshire, cleaner air 'than in any time in living memory', and climate-related deaths falling 'by about 90% throughout the past century', alongside apparently encouraging statistical developments (Nelson, 2023c, p. 18). Others use statistics on more global deaths arising from cold weather to argue that 'warming has meant fewer people dying' (Ridley 2022, p. 16), and Nelson adds that 'adaptation and tech give us the tools to handle extreme weather, [meaning] we have no reason to fear the summer' (Nelson, 2023b, p. 16). Such comments ignore the fact that heatwaves are a dangerous climate phenomenon for many people suffering from them—which tend to be black women in much poorer

[4] Some argue that it was introduced via secondary legislation with brief debate and 'no Parliamentary vote' (Grant, 2023, p. 15).

countries (Renwick et al. 2024a, b). Most recently, news channels such as GB News in August 2023 seem to have moved the UK down to a new and even lower level of the climate debate, in showing a video of two presenters laughing at their guest Alex de Koning describing the process of how the human body dies through 'wet bulb' like heatstroke, as him making 'ridiculous comments', and then asking him to 'grow up' (The London Economic, 2023).

Sunak's Future

Whatever the truth and accuracy are about the trends and taste in the UK climate debate above, Rishi Sunak's new stance on the climate may not produce (m)any new green jobs, as he seems to be repeating the same policy mistakes or disasters (cf. Dunleavy, 1995) on green policy that former conservative UK leaders Cameron, May and Johnson all made, because there is plenty of time until the next election for the British media to unpick Sunak's new pro-fossil fuels, pro-motorist and pro-private jet approach. Further, Sunak's former energy deputy Shapps stressed his new department title of energy first (Nelson, 2023c, p. 18; Shapps, 2023b, p. 27), which Shapps's replacement Claire Coutinho has continued, and means its other title of net zero is perhaps a secondary priority. In doing so, Sunak may be on a political road to nowhere here, as UK voters do not like broken promises (Payne, 2023, p. 19), which arguably include ones Sunak and his Cabinet have already made on achieving net zero by 2050. Further, Sunak arguably gives the impression of being a mini-Donald Trump in treating Britain like his own personal fiefdom, as he uses words in his media interviews like: "I think/my interests/I am not prepared/my private flights" (above), which seem self-centred and perhaps non-green. Indeed, further evidence of UK government policy lying in the hands of single politicians is noted elsewhere, as Minister Gove decided to press ahead with a new UK coal mine and to deny leading retailer Marks & Spencer the ability to develop their asbestos-riddled flagship London Oxford Street store, which their Chief Executive described as a 'capricious and chilling' decision, riding 'roughshod over the facts and experts' (Machin, 2023, p. 23). Gove additionally announced that the British government 'will scrap EU-era pollution laws which prevent homes being built' (Oxley, 2023, p. 2). Thus, is some UK power-making now so concentrated that it seemingly lies in the hands of single conservative politicians like Rishi Sunak and Michael Gove?

A question arising then is whether, because of Sunak's approach to green policy probably is likely to lose him some votes, he is the least astute recent British political leader, or if Sunak is the smartest one in the room. Perhaps Sunak knows something big about UK climate change policy that other external commentators and observers do not, but he is simply not telling us, that his new low-cost type of greening may work. For example, is the 'cost' of running down the UK oil and gas sector in terms of lost government tax revenues and jobs too huge for Sunak to contemplate? Or is it that:

> The Tories facing obliteration at the next general election on current trends could yet turn the tables if they dump "net zero" and draw up some workable, realistic and popular alternatives. (Johnston, 2023, p. 14)

In turn, is it, as David Frost states, time for Sunak to:

> Delay the net zero 2050 target. Abolish the deadlines on boilers and EVs. Get fracking and build low-carbon modern gas power stations and zero-carbon nuclear. Stop wasting money on green levies, and if we must use renewables make them stand on their own two feet. (Frost, 2023c, p. 19)

If so, the legacy that such green policy changes is one that a new UK leader like Sir Keir Starmer might potentially have to deal with, and what then happens next to British green job creation becomes a consequence of such changes. Do we get more dither and delay on achieving net zero by 2050 from Sunak (Marr, 2023, p. 1), as we did under the successive conservative administrations of Cameron, May and Johnson? Here, stories are emerging that either Sunak could stick to his existing green commitments or he might delay or dilute them (Parker and Fisher, 2023, p. 15), car makers now 'fume' at his government's green 'dithering' (as they want stability and investment) (Mustoe, 2023, p. 18), and the next general election might delay decisions on the UK nuclear energy programme too (Oliver, 2023c, p. 33). Here, Labour's Starmer has already expressed his frustration with Sunak not seemingly cooperating with the EU over a response to Joe Biden's green plan (Boscia, 2023, p. 2) (although

Sunak is also reported to be talking to the EU about rules regarding electric car parts sourcing) (BBC News, 2023, pp. 1–3), and long delays in responses to consultations on net zero policies have become a feature of recent (Sunak-led) government policymaking (Riley-Smith et al. 2023, p. 4). Right now (summer 2023), the UK faces serious competition to attract new green innovation, 'as more ambition is needed' and Britain has to be 'more holistic' to gain 'the next wave of projects' and 'reward certain investments in return for equity' (Kapadia, 2023, p. 18). As Sunak's Treasury under Chancellor Jeremy Hunt 'has pretty much ruled out attempting to replicate Biden's IRA, and his strategy has instead 'been to try and tap into America's green largesse' (Warner, 2023b, p. 26), Sunak's Tories may need to change and become a 'pro-consumer party – anti-inflation, and pro-driver' to win (Stanley, 2023, p. 16). As such, Sunak's political judgement regarding his approach to green policies including new, decent green jobs may need sharpening if he wishes to win the next British general election.

Sunak's Late Summer 2023 Reshuffle

In early September 2023, Rishi Sunak's reshuffle saw him place his close friend in politics Claire Coutinho MP as Energy Secretary, which perhaps heralded a subtle change of policy from the government away from green causes, and their 'huge costs' for 'ordinary families' (Martin et al. 2023, p. 5). Coutinho is a CEN member, has previously championed environmental issues like growing wild rural spaces and protecting the green belt, and agrees with Sunak to protect UK households from huge financial costs. She is not opposed to net zero, has pledged to only shop for second-hand clothes and use more public transport, and held a green surgery with voters. She said: "I will work with the PM to safeguard our energy security, reduce bills for families, and build cleaner, cheaper homegrown energy" (in Martin et al., 2023, p. 5). Here, Sunak is 'choosing his last few pre-election battles, gathering and promoting his closest allies and getting ready for his final push' (Nelson, 2023c, p. 14), which includes a 'face down' with British UK government climate advisers to halt airport expansion, and instead expand them, while hoping rapid development of green aviation fuels can 'decarbonize the sector' (Malnick, 2023c, p. 1).

Labour's Green Plan

In terms of alternative British green job plans, Labour's new policy response over the summer of 2023 led by Shadow Energy Secretary Ed Miliband seems worth highlighting, not least because Labour may be one of the few parties to either win the next UK general election outright or become a major player or coalition partner in it from 2024 to 2025 (cf. Nelson, 2023d, p. 14), meaning Labour may partly dictate future British green jobs policy. Here, Labour states that they will look to provide '100,000 jobs in the offshore and onshore wind; commit to make Britain a clean energy superpower by 2030; set up GB energy, a publicly owned company; create a National Wealth Fund to secure good jobs; and insulate 19 million UK homes over a decade' (Miliband, 2022, p. 5). Moreover, Labour previously committed to spending £28 billion every year on green investment (Murray, 2022, p. 1), which is higher than the conservative's original £9–12 billion per year[5] (Renwick, 2022, 2023). But Labour then lowered this figure down to £15.38 billion per year until 2030[6] and promised "good, union jobs" from their Green Energy Plan, which would involve 'wind farm subsidies, home insulation, electric vehicle battery factories and a new generation of nuclear power stations' (Oliver and Lawford, 2023, p. 16). Labour 'tax and spend plans' were also reduced to only being enacted in the second half of any first Labour Parliament term (i.e. 'fiscal orthodoxy' is applied), because Labour argues Britain's public debt is already so high they do not want to raise taxes to go green. Whichever exact financial investment figure emerges, Labour is 'committed to creating a million good, new jobs from it' (Lawson, 2023,

[5] These figures differ, as one was an original UK conservative estimate under the Boris Johnson government (£9 billion) and another shortly after his exit (£12 billion). The latter, higher figure, was one former Energy Minister Shapps quoted as part of their £100 billion green energy spending from mid-2023 until 2030 (on BBC Newsnight, 2023b).

[6] As Shadow Chancellor Rachael Reeves said £28 billion per year was a figure Labour could not now commit to, because of economic uncertainty (Maidment, 2023, p. 8).

p. 24). Here, Labour's Thangam Debbonaire said: "Sunak is taking us backwards, Labour will honour 100 new oil and gas licences, but we want four times more onshore wind too. The Tories say either growth or green, but we need both, and we need a rapid transition" (on BBC Newsnight, 2023b), and Labour leader Starmer's approach is simple: he wants to "getting Britain thinking big again" (Starmer, 2023, p. 5).

Personally, Labour leader Starmer—just like conservative leader Sunak—at first sight also seems on board with publicly adopting green jobs policies, backs onshore UK wind power and supports worker 'just transition' initiatives. Indeed, reports are that Starmer 'will scrap the ban on new onshore wind farms, tear up planning laws, and enact a strategic plan for skills to double onshore wind output by 2030 to create British power for British jobs' (Gutteridge et al. 2023, pp. 1, 5), and Labour has pledged £3 billion to decarbonize UK steelmaking and make large investments in renewables too (Jolly, 2023, p. 53). Further, industry bosses at Centrica (who own British Gas) and the GMB trade union argue that 'the UK's net zero economy could be worth £1 trillion by 2030' and 'create tens of thousands of new, well-paid, skilled jobs right across the UK' if a '10-year skills revolution' is enacted. Yet this is a huge challenge for any possible incoming Labour government to deliver, as the UK workforce is ageing and it has few apprentices and not enough vocational training. Further, any British government needs to proportionally match US and EU commitments to fund this green transition or risk being left behind, with a clear strategy and plans to develop renewables, hydrogen, mini-nuclear reactors and CCUS (O'Shea, 2023, p. 21). Nonetheless, others see that Starmer's Labour have such a plan, with their "fairer, greener future" slogan, and 'even bigger green investment per capita than Joe Biden's' green plan, 'meaning sticking to its green guns is Labour's path to victory' (Toynbee, 2023, pp. 1–2). Here, Shadow Chancellor Rachael Reeves states that:

> The push for net zero by 2050 should not make people poorer, but we need to stick to the targets because businesses need certainty, and businesses and local people know what is best for them; this government is piecemeal; we need eight battery factories to cement our position as a serious car manufacturer; it is about planning and skills. (in Malnick, 2023c, p. 6)

But again—and just like Sunak—Starmer's public comments also 'endorse longer sentences for people [climate protestors] who glue themselves to roads' (Monbiot, 2022, p. 3), and he was quoted as saying he "hates tree-huggers" (in Wheeler and Dahan, 2023, p. 1), a quote which was later revealed as 'a Tory lie, according to his [Starmer's] team' (Toynbee, 2023, p. 2). Starmer most recently emerged as being anti-Ulez expansion in July 2023, as it seemed to have cost his party votes in the Uxbridge by-election, and one unknown senior Labour figure states: 'People like the idea of green policies until they stop them doing something, or cost them money' (Savage et al. 2023a, b, p. 6).

Labour's green jobs policy needs more development and clarity, as they have big questions to answer on it, just as Sunak's conservatives do. Indeed, the UK arguably now has an emerging political consensus where Labour and the Conservatives tend to agree 'on almost everything' and where 'stopping climate change comes distant second to fiscal prudence', meaning that 'sweeping change' is not expected, but 'squabbling over a few billion pounds worth of annual state spending – mere crumbs off a £1.2 trillion cake will be' (The Economist 2023, p. 26).

Where Are We Now, and Where Are We Going?

From the points above, a clear dividing line now seems in place in British politics and green policy. Labour under Starmer appear to engage in a state-led, top-down plan to create green jobs and power the grid, while the conservatives under Sunak seem to prefer state intervention that supports 'transition' fossil fuels and some renewables and to rely on private sector funding, equity, markets and technological innovation to gain energy security and generating green jobs as a by-product. But a big risk for Sunak is what happens next if the financing and markets fail. With such competing green visions now known, the UK election battle lines on them will begin. Nonetheless, problems arise for all British politicians in winning over voters regarding tackling climate change, alongside opportunities emerging for them, as on the downside:

> "If going green costs money that not everyone has, then ultimately there are two plausible political responses. The first is utterly unconscionable since it means reneging on net zero. The second is to find the money for a genuinely fair transition, and fast. Unsurprisingly, most politicians are keener to talk about shiny new green jobs or cheap solar energy than about where they'll raise the money to fund new scrappage schemes and grants" (Hinsliff, 2023, p. 2).

On the upside for the UK conservatives, electoral boundary changes mean 585 of 650 Westminster seats are changing, with England gaining ten seats in the south-east and south-west, and 'electoral gravity' moving to mainly conservative voting suburbs and towns, which may benefit the conservatives by 5–10 seats come the election (The Guardian, 2023a, p. 2). Such gains may appear small now, but these fine margins may prove important for Sunak's conservatives later. However, a small frustration for them is that some of their traditional supporters at the right-wing publication *The Economist* seem less friendly because it says "Britain has been poorly governed for 13 years"; they argue that "the conservatives are passing on a rotten inheritance" to Labour and that 'the next period of chaotic conservative government is only a few years away' remains a political possibility (Beckett, 2023, p. 3).

Here, because Britain is currently the sixth largest economy globally (GOV.UK, 2023) what happens in it regarding green jobs matters, because if the UK (and other larger economies) do not lead the green jobs industrial revolution (which I imply herein that Britain is now not doing), then who will? And what does such UK leadership mean for developing economies looking to go green and protect the ecology too? Ergo, it seems vitally important for the UK to improve, so that it can build better eco-employment to benefit future generations to come. The task at hand for Rishi Sunak to now achieve a breakthrough in green policies (above) seems immense. As one commentator states:

> "This is not about individual politicians, nor is it even about the parties. Everything seems frozen by uncertainty, ignorance, and fear. The understanding is not there, the ideas are not there, and the ability of the state to devise and execute plans and policies is certainly not there. The challenges we face are truly enormous, yet our politics has never seemed smaller" (Timothy, 2023, p. 14).

Further, Sunak may need to detoxify the British national debate on green tech, as in the case of supposed UK 'national revulsion' against heat pumps, 'better, cleaner, hotter' ones are on the way using helium, ionocaloric technology and natural rubber, meaning British consumers may be tempted to buy more than the 40,000 heat pumps they purchased so far in 2023 and join the US, Poland, Norway, Sweden, Finland, France and Italy in installing more (Evans-Pritchard 2023, p. 21). Indeed, others see heat pumps working efficiently in very low temperatures in Finland and Canada and state UK ones work less efficiently 'because they are fitted poorly' (Alsop, 2023a, p. 1).

Sunak's Speech of 20th September 2023

Rishi Sunak's most recent (September 2023) speech on changes to UK government policy regarding net zero by 2050 seems like a watershed moment in UK politics and green jobs if the reaction to it is anything to go by. Rather than repeat or paraphrase Sunak's own words again, readers are guided to gain full details of his speech which appear at: https://www.gov.uk/government/speeches/pm-speech-on-net-zero-20-september-2023#:~:text=We%20will%20never%20impose%20these,in%20the%20same%20way%20again

Here, reaction to Sunak's speech falls into two main areas: fact-checks and problems emerging, which I now detail.

Fact-checks on Sunak's speech raised issues of some seeing Sunak sounding 'like an asset stripper. Absolutely unprepared to invest for the future' (Oldridge, 2023a, p. 1), practising 'deceit' (Oldridge, 2023b, p. 1) and 'wishful thinking' (Nevett, 2023, p. 1) plus sparking an 'industry backlash' (Brown, 2023, p. 1). Here 'business anger grows', as Ford Motor Company said the private sector requires "ambition, commitment and consistency" from the UK government, but Sunak's speech undermined all three (McKay and O'Leary, 2023, p. 1); 'banks hate' his U-turn as 'uncertainty and risk are expensive' (Dunn, 2023, p. 1). Here, as businesses require certainty and policymakers need consistency, Sunak's policy change seems like a problematic issue for them both. Others view him as practising 'bad economics dictated by cynical politics' (The Guardian,

2023b, p. 1), someone who is not fooling the UK media (Vince, 2023, p. 1) and 'turning the environment into a US-style political wedge issue… purely to score points. It is reprehensible' (Goldsmith, 2023, p. 1). Indeed, Sunak seems to be ignoring the fate of the last Australian government which backtracked on the climate and lost an election (Murray, 2023c, p. 1). Here, the creeping climate-scepticism outlined in *The Daily Telegraph* editorials and pieces (above) now seems partly realized under Sunak's regime, and he has divided his own party on achieving net zero too (BBC Newsnight, 2023c).

Sunak's speech may cause him new problems, as one moderate Tory said 'delaying the UK's ability to lead the energy transition will mean less investment, slower growth, fewer jobs and ultimately higher bills' (Skidmore, 2023a, p. 1), another Tory asking why he wants to 'shatter' the UK's 'broad, non-partisan consensus' on climate issues (Clarke 2023a, b, p. 1), and public opinion polls saying it makes the UK government look 'untrustworthy, reckless and backwards' (ECIU, 2023a, p. 1). Moreover, Tory MPs are very surprised about his announcements on pushing back petrol/diesel car bans, as they seem 'anti-business' and break a promise he made in private to them (Coates, 2023a, p. 1), that they 'will not help economically or electorally' (Sharma, 2023, p. 1), provide 'only a cost, not a benefit' (Skidmore, 2023a, p. 1), and other Tory MPs are considering submitting no confidence letters in Sunak too (Politics UK, 2023a, p. 1). Some commentators see his plan as 'a fundamental threat to lots of new private investment, as £6 billion is going into EV charging' (Hebditch, 2023, p. 1), and that it will cost Britain '£4.2 billion and 30,000 jobs' (Walker, 2023b, p. 1). Tech sector reaction is that Sunak now 'turns heads towards countries with more stable regulatory environments' (Vaughan 2023a, b, p. 1), that it offers him 'little electoral advantage (because it's not popular)' (Walker, 2023c, p. 1), as he has made a 'huge strategic mistake' (Skidmore, 2023b, p. 1), with '40% of people less likely to vote Tory now' (Politics UK, 2023b, p. 1), and that despite stating he remains committed to net zero by 2050, the risk for Sunak is that his 'nuance is lost or that voters simply don't buy it' (Balls, 2023, p. 1).

Other problems include whether Sunak's speech is his 'secret weapon or a gift to Labour?' (Aurora, 2023, p. 1), a form of 'cheap politicking'

(Nowak, 2023, p. 1), and 53% of 2019 Tory voters thought more spending on renewable energy would bring down living costs (ECIU, 2023b, p. 1). Further, the heat pump industry says Sunak is 'damaging investor, installer and consumer confidence' and that he has not provided evidence for some of his claims (Ralston, 2023a, p. 1), while voters want more green investment and 'not less' (Vinter and Stacey, 2023, p. 1), and Sunak's climate about-turn 'could risk the UK's legally binding emissions targets' (Peston, 2023, p. 1). The centre-green Tory CEN group argue for a 'more positive narrative' than Sunak's speech (Conservative Environment Network (CEN) 2023, p. 1); over 400 Non-Governmental Organizations (NGOs) including Nestle, Aviva and EON published an open letter 'to lay their deep concerns about any weakening of net zero policies' (E3G, 2023, p. 1), and the *Financial Times* labelled Sunak's speech as an 'unwise retreat on climate policies' (Financial Times (FT) 2023, p. 1). Key reactions 'from industry, academia, NGOs and many politicians' include words like 'pathetic, betrayal, stupid, chaotic and incoherent' (Polman, 2023, p. 1), and his speech 'may scarcely cut through to voters at all' because UK voters support green policies and net zero, they do not blame such policies for rising living costs and they see policy U-turns 'as a bad look' and are 'rolling their eyes' against politics (Frayne, 2023, pp. 1–3).

Some opinion polls suggest that 'two-thirds of the UK public do not trust the Tories on environment' (Slawson and Gecsoyler, 2023, p. 1), Sunak has risked Tory seats as a 'climate majority' exists in Blue Wall and key marginals (Stone, 2023, p. 1), and 'only 22% of Britons trust Sunak on the climate' (Horton, 2023, p. 1). Sunak is now seen as the problem because he has 'become the status quo' (Walker, 2023d, p. 1), his net favourability score has dropped to a new low of -45 (YouGov, 2023, p. 1), and 'it looks like the people love net zero after all' (Ralston, 2023b, p. 1). Indeed, while Sunak made his speech when most of his environmentalist colleagues were out of the country, UK regulator Ofgem chief Brearley says his 'economic case for pushing back net zero targets is "analytically very weak"' (in Kenyon, 2023, p. 5), while other international opinion is that Sunak's plan 'is bad news for his country and his planet' (Karl, 2023, p. 1). Other commentators argue that Sunak's changes 'will leave renters colder and poorer with sky-high bills' (ECIU, 2023c, p. 1; Middleton and Ahmed, 2023, p. 1). The largest UK conservative party donor in 2019 John Caudwell said he is "beyond shocked" at the "madness" of Sunak's reversal (in The Times, 2023b, p. 1), individual senior UK conservative MPs rail against Sunak as pursuing 'the worst kind of culture war' (Harvey, 2023, p. 1), and more conservative UK newspapers talk of Sunak's 'green gamble over net zero', e.g. *The Times* (2023c, p. 1), and Sunak has been warned he is 'likely to face legal challenges' over his changes (Harvey et al. 2023, p. 1). Yet while other conservative UK newspapers such as *The Sun, The Daily Mail* and *The Daily Telegraph* all support Sunak over his net zero changes, they are also seen as 'preparing the Tories' downfall' over it, because they also backed the short-lived Liz Truss regime which spooked the bond markets and crashed the Pound Sterling (Elledge, 2023, pp. 1–3). Nonetheless, other UK conservative reporters argue that Sunak is adopting 'a bright green agenda focusing on tech and green progress' and 'joining a European trend towards more pragmatic green policies' (Nelson, 2023e, p. 1).

Of course, Sunak operates in a wider context which includes several aspects, namely, managing his external and internal pressures; negating his traditional supporters; and bypassing opposition to his plans, which I now illustrate.

Here, Al Gore sees Sunak as under pressure from external sources, as someone "captured" by fossil fuel companies (Gore in Mahmood, 2023b, p. 1), yet he also gained praise from Donald Trump who said his net zero changes were "smart" (Gore in Sky News 2023a, b, c, p. 2). Sunak is under pressure internally in his party, by right-wing MPs wanting a 'debate' on decarbonization costs (Conservative Home, 2023, p. 1) to avoid 'bankrupting Britons' (Braverman in Baker, 2023, p. 1), and for him to reduce green spending and provide more personal income and inheritance tax cuts too (ECIU, 2023d, p. 1). Sunak himself wants to 'do something he believes will go down well with Tory voters' (Coates, 2023b, p. 1) and 'grab the agenda' (Mason, 2023, p. 1). To manage such internal pressures, Sunak appointed the chief economist of Shell to run the UK climate watchdog CCC (Oldridge, 2023b, p. 1), and new Energy

Secretary Claire Coutinho[7] said that 'it is wrong to expect people to "change their lives" or "punish" people to reach net zero (in Ferguson, 2023, p. 2). Additionally, Sunak granted the first wave of 21 North Sea carbon storage licences (Murray, 2023d, p. 1) and has overseen his transport department 'weaken their already weak climate commitments' (Anderson, 2023, p. 1), while offering up a lot of straw men to criticize, which 'simply were not' UK government policy (e.g. banning flying, taxing meat and placing trash into seven different bins) (Clarke 2023a, b, p. 1).

Sunak seems to reject traditional UK conservative supporters, as his green changes saw one commentator argue that "business generally has pretty much lost it with this government" (Coates, 2023b, p. 1), and multinationals are instead calling for a 'tripling of renewable power capacity by 2030' (Keating, 2023, p. 1). UK public opinion seems mixed on Sunak's new approach to net zero, as 54% of Britons prefer tax cuts to be focused on renewables (ECIU, 2023d, p. 1), 46% said the government should stick to its (prior, original) plans (ECIU, 2023e, p. 1), and UK businesses and voters appear keen on electric vehicles (EVs), as Britain has the second highest rate of EV sales for major economies (ECIU, 2023e, p. 1). Here, Sunak's announcement was 'about allowing fossil fuel companies to continue fleecing us' (Oldridge, 2023a, p. 1), and he uses 'the tactics of doubt, delay and distraction devised by none other than... Big Oil' (Shukman, 2023, p. 1).

Sunak's emphasis on cost-control regarding net zero seems to ignore the jobs, legal and moral case for advancing the green concern and seems similar to his bypassing social issues like UK National Health Service (NHS) outcomes, where the waiting list hit 7.68 million and from 2018 to 2022, 1.5 million people died waiting for hospital treatment (Sikka, 2023a, p. 1). Sunak's net zero changes also seem contradictory, as he is reducing UK climate commitments and leading the way in fighting climate change (Griffiths, 2023, p. 1), he has disbanded a task force to speed up home insulation and boiler upgrades (which only formed in March 2023) (Wells, 2023, p. 2), is lambasted for building 'no new

[7] Who herself is linked to a policy group funded by ExxonMobil which targeted climate activists (Lott-Lavigna, 2023, p. 1).

industries' (Sikka, 2023b, p. 1) and 'has handed Tata Steel a £500 million green subsidy to fire 3,000 steelworkers' (Sikka, 2023c, p. 1).

Ergo, is Sunak very right-wing and more focused on financial management within green policy and less so on social conscience issues? This question arises because instead of Sunak leading on climate, it seems that his former UK Net Zero Review chief Chris Skidmore MP does, as Skidmore launched an international climate declaration to secure cross-party and cross-country support for climate action (McKeon, 2023, pp. 1–3), while Sunak acted secretly and 'blindsided' Whitehall and his own UK delegation in not telling them of the extent of his net zero changes while they were away at a New York climate summit (Coates, 2023a, p. 1). Thus, Sunak seems trapped by the centrifugal force of conservative political ideas, traditions and history, as his only answers to the challenge of meeting net zero by 2050 seem to lie in giving more support to the UK-based fossil fuel industry, i.e. a no-change/status quo scenario (cf. Goodwin, 2014). Indeed, it could be argued that while the old UK conservatives seemingly wanted to ignore or bypass trade unions, Sunak's new ones appear to want to do the same to business. Sunak's dilemma could be whether his new approach to net zero will win him more votes, and how he can also build decent, new UK-based green jobs in future. Right now, his conversation and narrative have moved from a focus on green jobs to one on green politics, meaning that the politics of green job creation (or neglect) looks set to take over mainstream UK debate. Ergo nowadays, due to Sunak's net zero U-turn above, the GJT era seems like a distant memory.

Study Limitations

Several limitations apply to this study. These include it only focusing on a single country and one national context, alongside analysing and discussing a single British conservative/Tory political party. It is also limited by only examining one key UK government report in depth, the GJT, and critiquing it through the single theoretical lens of conservative political ideas. Further, this study has the limitation of not robustly engaging with right-of-centre British lobbying groups like Net Zero Watch (NZW) (due to issues around the accuracy of their information), and not critically investigating them or the GJT using liberal or left-wing political ideas and alternative theoretical lenses. Additionally, this monograph could be seen as somewhat Anglo-centric, by focusing on the UK, EU and US at the expense of less focus on the BRIC states (Brazil, Russia, India and China) (see Razzouk, 2023, p. 1; Murray, 2023e, p. 1 for details on China's climate actions and future plans), and other non-BRIC countries, alongside less emphasis on the original UK role in creating the climate emergency itself (see Stop Cambo, 2023, p. 1). Still, to counter this Anglo-led limitation, I offer a view of green HRM from the Global South countries elsewhere (Yusoff and Renwick, 2024), which includes some discussion of green jobs, and a critique of them as a North-Western phenomenon (Renwick et al. 2024a, b), I have undertaken in-depth work on green HRM in Brazil (Renwick, 2009; da Silva Souza et al. 2014; Renwick et al. 2024a, b), and I am now completing a 12-country analysis of green jobs among non-Western countries including all four BRIC states too (Renwick, 2024). Further, readers can consult the international press and media for stories on the UK's role in driving the climate crisis.

Moreover, the benefits of the approach I adopted herein remain, including critically analysing UK green jobs policy using a new theoretical lens in conservative political ideas and providing an in-depth, reflective analysis of the GJT. Further, this study critically highlights gaps in British political and business decision-making, the lack of employee voice arising, and that the GJT places less priority on skills and training too. To my knowledge, this monograph is the world's first in-depth analysis and critique of the GJT; it provides an alternative and different road map ahead on the issue of UK green job generation and aims to act as a key learning resource to help increase knowledge on British-based green jobs and as a central focus in helping to create such decent UK jobs too.

Policy and Practitioner Implications

UK policymakers and politicians might wish to formulate an agreed position on green policies and to stick to them, to think through how to fund such initiatives and fully organize them, to follow through on such policies, and try not to let current events reshape them too much and only let minimal policy drift in (cf. Hogwood and Gunn 1984). Further, policymakers could ensure equal access to new labour market opportunities in less represented regions and demographic groups such as women and younger people, alongside 'training for skills programmes and targeted transition policies and programmes' which detail clear career paths at different points in the transition (Valero et al., 2021, p. 7). Additionally, a key practitioner implication is that it seems time to move away from purely focusing on green job numbers (albeit recognizing their importance in building a new green economy) and embrace issues around staff sustainability skills, as they need building up among British workers.

Future Research Ideas

Scholars could initially draw inspiration from work done in the already-established fields of Political Corporate Social Responsibility (CSR) and the Management of Socio-Political Issues and research the aspects detailed therein (see Al-Esia et al. 2023, p. 19; Sun et al. 2023, pp. 23–24), alongside establishing a new HRM sub-field which I title Political HRM. Here, Political HRM may be understood as an umbrella term containing a broad church of differing (un)orthodox, mainstream and critical viewpoints and perspectives, to stress the importance of politics as a key factor or variable in researching external contexts influencing and interacting with organizations and their staff in HRM studies. Here, at the macro-level, researchers might usefully assess democratic versus authoritarian political regimes, and varieties of capitalism (VoC), to assess their impact on shaping respective national government green policies and how such contexts translate into various recipes of green job creation or loss in different countries.

At the meso-level, scholars could investigate the role of the state, the role of the individual and personal liberty issues, and a focus on politics, policy and politicking (broadly defined) and any possible interactions with HRM, plus links between the intersection of government regulation and business level operations in HRM, e.g. in connecting with and shaping corporate HR strategies, policies and procedures and workplace HR practices, and vice-versa. Here, green examples could include how national government green jobs policies and plans shape firm-level, organizational and departmental HRM systems, policies, procedures and practices, the role of HR managers, line managers and employees in this process, and issues of training and skills for workers undergoing the transition from fossil fuel (non-green) to renewable energy (greener) jobs, alongside staff inclusion and exclusion in doing so. Research could also focus on any policy mistakes or disasters (Dunleavy, 1995) and real-world policy analysis (Hogwood and Gunn 1984) issues arising regarding such green jobs policies. Further, researchers may use different political ideologies to critically analyse new green job data, i.e. from Goodwin (2014) as detailed herein, and other political theoretical lenses such as liberalism, socialism, Marxism, anarchism, authoritarianism, feminism, nationalism and green ideologies, plus neo-institutionalism, political economy, post-colonialism and neo-Schumpeterian approaches too (see Adler et al. 2021).

Moreover, future research at the micro-level might examine the fate of the different political leaders involved in pursuing green jobs policies. For example, will Rishi Sunak become a new James Callaghan and perhaps 'find that a popular leader cannot compensate for an unpopular party'? (New Statesman, 2023, p. 1). Will Labour leader Keir Starmer appear as a new version of the 'inexperienced yet highly confident' David Cameron (Jacobs, 2023, p. 16) or Neil Kinnock (Kavanagh, 2023, p. 1)? Can Starmer solve the historical, systemic policy and governance issues involved in UK green job investment and creation, while doing similar regarding the British health service, social care, industry, Brexit and immigration, while not raising UK taxes too? Will Boris Johnson make a political comeback as other UK conservative leader returnees did (e.g. Sir Winston Churchill, William E. Gladstone and Harold Wilson), or will Liz Truss with her 'pro-growth' agenda, or someone else, as 'Tory women jostle for position', i.e. Kemi Badenoch, Suella Braverman, Priti Patel and Penny Mordaunt (The Daily Telegraph, 2023n, p. 6)? As the mental state of the UK conservative party is so changeable and fluid, and it previously contained arguably uninfluential leaders like Sir William Hague and Michael Howard, they could choose any leader candidate. Certainly, researching such political leadership issues and how they pertain to, and link to, green jobs policies perhaps seem a useful and interesting line of research to pursue.

Other research avenues could include investigating worker attitudes, experiences and beliefs on decent, green jobs (or a lack thereof), how they perceive such roles and how such contemporary insights compare. Here, scholars could assess local employee workplace conversations, interactions, briefing groups and encounters which shed light on the opportunities, insights and challenges arising on decent, green jobs. Moreover, future qualitative-based research could reveal findings from stakeholders co-creating value, vignettes and ethnographies to reveal eco-employment as a lived experience too. Here, scholarly interrogations could focus on structural and systemic inequalities, the regulatory framework, a lack of political will and other barriers preventing local worker take-up of decent, green job roles.

Conclusions

Overall, discursive approaches like the work herein analysing the British GJT text possess value in revealing some abstract and eluding aspects of politics, organizations and society, e.g. factors like power, identity, conflict and ideology (Oswick, 2012, p. 488). Here, in terms of UK green employment initiatives, funding and retraining issues seem lost amongst all the political talk and (in)action, while grandstanding (Tosi and Warmke, 2020) becomes the main way the achievement of British government goals and socially useful outcomes such as new, decent green jobs dominate, are publicized, and disseminated to the wider public (cf. Oswick, 2012, p. 487). Such political talk, inaction and grandstanding are seen in the many arguably hyperbolic examples detailed in earlier chapters herein. Moreover, it is not clear how British workers can gain new green jobs if such opportunities are not properly funded, and how staff employed in the polluter industries can transition to green jobs if little retraining and reskilling exists for them to do so. As such, the critical analysis undertaken here seems warranted, as it reveals how and why UK PM Rishi Sunak's position on green jobs (and those of his predecessor Tory leaders) is perhaps beguiling, as while he sounds green (his language), his actions seem pro-fossil fuels (his language-in-use) too (cf. Oswick, 2012).

In concluding this research monograph, I finish on a relatively optimistic note. Despite my many, extensive criticisms of the GJT, the British conservative political party, and their ideas on UK green job policy initiatives (above), the stimulus towards a 'green industrial revolution' by the often-derided former conservative leader Theresa May does show something of a start towards creating a green jobs future for the British economy and one pathway out and away from historic UK decline. Indeed, if enacted more fully, scope exists for it to create something of a brighter, greener future for British workers, if UK Labour politicians can now enhance and further expand it (see Valero et al., 2021, p. 4). I am also optimistic because of: the falling costs of and available incentives around renewable energy, growth in British citizen movements demanding action regarding the climate crisis, UK and international business uptake of green job creation, British Universities embedding sustainability into their curricula, global government work on tackling, and media coverage of the climate crisis, public opinion data that UK citizens wish to see more government action to tackle the climate emergency, and the power of government to enact macro-level policies (Stanley 2023, p. 16), bypass critics and reduce barriers to change (Elliott, 2023, p. 29).

Yet a key policy lesson specifically arising for British conservative politicians from this research book, and ample scope for some pessimism too, is that they need to provide more leadership to support and achieve net zero and be less fixated on, and obsessed about, their own conservative political ideology and ideas (especially the role of the state, society and the individual). Due to the deep, systemic faults in their approach towards building a new green jobs policy via their GJT (detailed herein), which are ideological, and way more historically long-standing than purely arising from Brexit (although Brexit has now exacerbated such faults), we arguably need more policy from conservative UK leaders and politicians in future and not more politics or politicking. As such, it may perhaps be time for them to make progress through using more typical, practical conservative tools like more rapid yet incremental change, and trial and error, to take a risk, properly fund UK green industrial policy, work harder on exactly how to implement green policy goals via state intervention, and use the public sector more widely (as the Americans and Europeans do), rather than relying on the private sector and private equity to do so. In making these points, I hope that this research monograph helps as a conversation starter to at least discuss British green jobs and industrial policy more widely and, through the critical angle I employ herein, to kick-start improvements in it that might ultimately play a part in providing some benefit for UK citizens looking to work in a green job in future.

References

Adler, P. S., Jermier, J. M., & Nyberg, D. (2021). '37th EGOS Colloquium: Organizing for an Inclusive Society: Meanings, Motivations, and Mechanisms, Sub-theme 15: [SWG] Organization Studies in the Anthropocene: System Change, Not Climate Change'. *Vrije Universiteit Amsterdam (VU), The Netherlands,* 1–3, Accessed 8th–10th July 2021. https://www.egosnet.org/jart/prj3/egos/main.jart?rel=de&reserve-mode=active&content-id=1610525130808&subtheme_id=1591864212381

Al-Esia, Z., Crane, A., & Iatridis, K. (2023). Political CSR and populism: Toward an information-based theory of political CSR. *Business & Society,* 1–26., First published online May 1, 2023, Accessed 29th September 2023. https://doi.org/10.1177/00076503231168078

Alsop, J. (2023a). *Twitter/X,* Tuesday 12th September, p. 1.

Anderson, K. (2023). *Twitter/X,* Tuesday 19th September, p. 1.

Atkins, E. (2023). *Twitter/X,* 31st July, p. 1.

Aurora, M. (2023). 'Net zero: Sunak's secret weapon or a gift to Labour?', *Sky News,* Thursday 21st September, pp. 1–9.

Baker, T. (2023). 'We're not going to save the planet by bankrupting Britons,' says Braverman – as car makers hit out at government', *Sky News,* Wednesday 20th September, pp. 1–9.

Balls, K. (2023) 'Opinion: Inside the Tory climate soap opera this week: "The WhatsApp groups are a bloodbath"', *The Guardian,* Thursday 21st September, pp. 1–6.

BBC News. (2017). 'New diesel and petrol vehicles to be banned from 2040 in the UK', *BBC News,* 26th July, pp. 1–6.

BBC News. (2023). 'Rishi Sunak talking to EU over threat to UK electric cars', *BBC News,* 18th May, pp. 1–9, https://www.bbc.co.uk/news/business-65631167
BBC Newsnight. (2023a). 10th June.
BBC Newsnight. (2023b). 31st July.
BBC Newsnight. (2023c). 30th September.
Beckett, A. (2023). 'Even the free-market bible is now turning on the Tories', *The Guardian,* Friday 21st July, p. 3.
Bell, A. (2022). *Twitter/X,* 26th October, p. 1.
Bennett, N. (2022) *Twitter/X,* 11th November, p. 1.
Bet, M. (2021). 'Boris Johnson warned Steve Baker could be biggest 2024 election challenge', *Daily Express,* pp. 1–6, published 5th June 2021, accessed 7th June 2021, https://www.express.co.uk/news/uk/1444702/boris-johnson-news-steve-baker-election-challenge-gas-boilers-spt
Boscia, S. (2023). 'Labour's Starmer slams lack of UK-EU cooperation over Biden's Inflation Reduction Act', *POLITICO,* 27th February, pp. 1–12, https://www.politico.eu/article/starmer-slams-lack-uk-eu-cooperation-responding-to-bidens-inflation-reduction-act-criticized-protectionist-london-labour/
Brown, F. (2023). 'Sunak accused of making "false argument" on net zero as plans spark industry backlash', *Sky News,* Thursday 21st September, pp. 1–8.
Chan, S. P., & Oliver, M. (2023). 'Wind farm developers risk leaving North Sea', *The Daily Telegraph,* Sunday 25th June, p. 27.
Clarke, S. (2023a). *Twitter/X,* Wednesday 20th September, p. 1.
Clarke, S. (2023b). *Twitter/X,* Tuesday 19th September, p. 1.
Coates, S. (2023a). *Twitter/X,* Tuesday 19th September, p. 1.
Coates, S. (2023b). 'Targets designed to drive net-zero look like they're about to be softened', *Sky News,* Tuesday 19th September, pp. 1–7.
Conservative Environment Network (CEN). (2023). Twitter, Thursday 21st September, p. 1.
Conservative Home. (2023). 'Decarbonising the UK could cost £3 trillion. We need to have a debate – Mackinlay', *Conservative Home,* 24th September, pp. 1–3.
Crerar, P. (2023). 'Rishi Sunak signals he could abandon green policies that cost consumers', *The Guardian,* Monday 24th July, pp. 1–6, at: https://www.theguardian.com/politics/2023/jul/24/rishi-sunak-suggests-delay-or-abandon-green-net-zero-pledges

da Silva Souza, C. P., Roglio, K. D. D., Renwick, D. W. S., & Takahashi, A. R. W. (2014). Contemporary trends in Brazilian human resource management studies over the last decade (2001-2010). *International Business Research, 7*(8) July, 29–46.

Daley, J. (2023a). 'Labour, too, is experiencing an existential crisis of purpose', *The Daily Telegraph,* Sunday 11th June, p. 20.

Daley, J. (2023b). 'Climate hysteria is a serious threat to mankind's survival', *The Daily Telegraph,* Sunday 30th July, p. 20.

Diver, T., Gibbons, A., & Leather, J. (2023). 'Starmer faces party mutiny over Khan's Ulez expansion', *The Daily Telegraph,* Thursday 6th July, p. 6.

Dunleavy, P. (1995). Policy disasters: Explaining the UK's record. *Public Policy Admin, Summer/June, 10*(2), 52–70.

Dunn, W. (2023). 'Banks hate Rishi Sunak's U-turn on net zero', *The New Statesman,* Saturday 23rd September, pp. 1–8.

E3G. (2023). Business and civil society lay out "deep concerns" at UK PM's plan to weaken net zero policies, *E3G News,* 23rd September, pp. 1–4.

ECIU. (2023a). *Twitter/X,* Sunday 24th September, p. 1.

ECIU. (2023b). *Twitter/X,* Thursday 21st September, p. 1.

ECIU. (2023c). *Twitter/X,* Saturday 23rd September, p. 1.

ECIU. (2023d). *Twitter/X,* Saturday 16th September, p. 1.

ECIU. (2023e). *Twitter/X,* Wednesday 20th September, p. 1.

Electrical Review. (2023). 'Spring Budget 2023: CCUS could receive £20 billion investment', *Electrical Review,* March 20th 2023, pp. 1–2, https://electricalreview.co.uk/2023/03/20/spring-budget-2023-ccus-could-receive-20-billion-investment/

Elgot, J. (2023). 'Boris Johnson allies behind bid to "take back control" of Tory party', *The Guardian,* Wednesday 4th January, pp. 1–2.

Elledge, J. (2023). 'Right-wing papers are preparing the Tories' downfall over net zero', *The New Statesman,* 23rd September, pp. 1–8.

Elliott, L. (2023). 'Climate: How leaders could save Earth… and why they won't', *The Guardian,* Monday 31st July, p. 29.

Evans-Pritchard, A. (2023). 'Heat pump tech will soon be hot enough to tempt you', *The Daily Telegraph,* Wednesday 23rd August, p. 21.

Ferguson, K. (2023). 'BILLS BID: New Energy Secretary Claire Coutinho promises not to "punish" families with higher bills to hit green targets', *The Sun,* 16th September, pp. 1–9.

Financial Times (FT). (2023). 'Rishi Sunak's unwise retreat on climate policies', *Financial Times,* 23rd September, p. 1.

Frayne, J. (2023). 'Sunak's Net Zero speech. It may scarcely cut through to voters at all outside the bubble', *Conservative Home,* 22nd September, pp. 1–4.
Frost, D. (2023a). 'The Tories should embrace national conservatism', *The Daily Telegraph,* Friday 19th May, p. 19.
Frost, D. (2023b) 'We Tories must get our act together or the party is doomed to lose power', *the daily telegraph,* Saturday 22nd July, p. 5.
Frost, D. (2023c). 'Ten policies to save Rishi Sunak from oblivion', *The Daily Telegraph,* Friday 23rd June, p. 19.
Gatten, E. (2023a). Boiler ban may lead to a shortage of electricity for new homes, *The Daily Telegraph,* Saturday 29th July, p. 4.
Gatten, E. (2023b). 'Net zero risks blackouts, ministers warned', *The Daily Telegraph,* Wednesday 21st June, p. 8.
Gatten, E., & Gutteridge, N. (2023). 'Net zero drive will add £120 to energy bills', *The Daily Telegraph,* Friday 19th May, p. 1.
Gibbons, A. (2023). 'Britons too hard-up to worry about climate', *The Daily Telegraph,* Sunday 3rd September, p. 6.
Goldsmith, Z. (2023). *Twitter/X,* Thursday 21st September, p. 1.
Goodwin, B. (2014). *Using political ideas* (6th ed., pp. 1–480). John Wiley & Sons Ltd.
GOV.UK. (2023). 'Research and analysis: Global trade outlook – February 2023', *Department for Business & Trade/Department for International Trade,* 28th February 2023, pp. 1–76, accessed 1st September 2023, https://www.gov.uk/government/publications/global-trade-outlook-february-2023-report
Grant, M. (2023). 'Net zero hardliners don't know their history', *The Daily Telegraph,* Wednesday 2nd August, p. 15.
Griffiths, A. (2023). *LinkedIn,* 23rd September, p. 1.
Gutteridge, N. (2023). 'Starmer "in bed" with Just Stop Oil backers', *The Daily Telegraph,* Monday 29th May, p. 4.
Gutteridge, N., Martin, D., & Oliver, M. (2023). 'Starmer: I'll push onshore wind to hot net zero', *The Daily Telegraph,* Monday 19th June, pp. 1, 5.
Harris, T. (2023). 'Scotland is now the testing ground for net-zero madness', *The Daily Telegraph,* Sunday 25th June, p. 23.
Harvey, F. (2023). 'The worst kind of culture war: Tories attack Rishi Sunak's reversal on net zero', *The Guardian,* Saturday 23rd September, pp. 1–3.
Harvey, F., & Cooke, P. (2023) 'Gas boiler lobby trying to delay heat pump plans, leaked papers reveal', *The Guardian,* Friday 21st July, p. 2.
Harvey, F., Horton, H., Niranjan, A., & Allegretti, A. (2023). 'Rishi Sunak likely to face legal challenges over net zero U-turn', *The Guardian,* 21st September, pp. 1–3.

Heath, A. (2023a). 'Thirteen years of Tory failure have shifted Britain radically to the Left', *The Daily Telegraph,* Thursday 6th July, p. 16.

Hebditch, R. (2023). *Twitter/X,* Tuesday 19th September, p. 1.

Helm, T., & Harvey, F. (2023). 'Leading energy companies to warn Sunak: "Don't back off green agenda"', *The Observer,* 16th July, p. 15.

Hinsliff, G. (2023). 'You cannot sell a green vision to those who cannot afford it', *The Guardian,* Friday 21st July, pp. 1–2.

Hogwood, B. W., & Gunn, L. A. (1984). *Policy analysis for the real world* (1st ed., pp. 1–289). Oxford University Press.

Horton, H. (2023) 'Only 22% of Britons trust Sunak on climate, finds Guardian poll', *The Guardian,* Saturday 23rd September, pp. 1–3.

Howard, E. (2015). 'Defra hit by largest budget cuts of any UK government department, analysis shows', *The Guardian,* Wednesday 11th November, pp. 1–2.

Ipsos.com. (2020). 'Two-thirds of Britons believe Climate Change as serious as Coronavirus and majority want Climate prioritised in economic recovery', *Ipsos,* Accessed 22nd April, https://www.ipsos.com/ipsos-mori/en-uk/two-thirds-britons-believe-climate-change-serious-coronavirus-and-majority-want-climate-prioritised

Jacobs, S. (2023). 'Comment: Keir Starmer isn't the new Blair: he's Labour's answer to David Cameron', *The Daily Telegraph,* Tuesday 11th July, p. 16, https://www.telegraph.co.uk/news/2023/07/10/keir-starmer-isnt-the-new-blair-hes-labours-david-cameron/

Johnston, P. (2023). 'Eco-fanatics forced me to sell my car – it might not even help the planet', *The Daily Telegraph,* Wednesday 21st June, p. 14.

Jolly, J. (2023). 'Welsh steelworks at breaking point over the cost of cleaning up its act', *The Observer,* 16th July, p. 52.

Kapadia, C. (2023). 'Sunak must be ambitious and heed Biden's investment blitz on energy', *The Daily Telegraph,* Tuesday 13th June, p. 18.

Karl, J. (2023). 'Rishi Sunak Shouldn't Pretend He's Doing Brits a Favor', *Bloomberg,* 21st September, pp. 1–7.

Kavanagh, T. (2023). 'No ideas, no plan, and no answers – Keir Starmer is the new Neil Kinnock and voters will see through his Sir Shifty act', *The Sun,* 16th July, pp. 1–4, https://www.thesun.co.uk/news/23075650/keir-starmer-new-neil-kinnock-voters-see-through-him/

Keating, C. (2023). 'Quantum leap in climate action: Multinationals call for tripling of renewable power capacity by 2030', *Business Green,* 18th September, pp. 1–8.

Kenyon, M. (2023). 'Sunak's net zero U-turn break with international opinion', *The New Statesman,* 22nd September, pp. 1–9.
Lawson, D. (2023). 'Keir Starmer's "growth" story is just a fairytale', *The Sunday Times,* 23rd July, p. 24.
Leake, J. (2023). 'Britain needs green fund to rival Biden's', *The Daily Telegraph,* Saturday 23rd August, p. 18.
Logan, R. (2017). '"Dear Donald" – MEPs pen "love letter" to President Trump to scrap climate change policy', *Daily Express,* 27th April, pp. 1–3.
Long, C. (2023). 'Comment', *The Sunday Times,* 30th July, p. 23.
Lott-Lavigna, R. (2023). 'Net zero minister linked to oil-funded group that targeted climate protesters', *openDemocracy,* 1st September, pp. 1–10.
Machin, S. (2023). 'Gove's capricious decision will have a chilling effect', *The Daily Telegraph,* Friday 21st July, p. 23.
MacKinlay, C. (2023). 'Scrap the ban! Growing calls to delay 2030 deadline for petrol and diesel vehicles', *Daily Express,* Tuesday 25th July, p. 6.
Mahmood, B. (2023a). 'How the Sun's anti-net zero campaign is linked to a lobby group funded by the freight and haulage industry', *Left Foot Forward,* pp. 1–4. https://leftfootforward.org/2023/08/how-the-suns-anti-net-zero-campaign-is-linked-to-a-lobby-group-funded-by-the-freight-and-haulage-industry/
Mahmood, B. (2023b). 'Al Gore blasts Rishi Sunak for net zero U-turn and suggests he's being captured by fossil fuel companies', Left Foot Forward, 21st September, pp. 1–4.
Maidment, J. (2023). 'Labour's green tech and jobs plan in tatters', *The Daily Telegraph,* Saturday 10th June, p. 8.
Malnick, E. (2023a). 'I am on motorists side, says PM as he orders review of anti-car schemes', *The Daily Telegraph,* Sunday 30th July, p. 1.
Malnick, E. (2023b). 'Gove: net zero can't become a crusade', *The Daily Telegraph,* Sunday 23rd July, p. 1.
Malnick, E. (2023c). 'Sunak defies net zero ban on new airports', *The Daily Telegraph,* Sunday 3rd September, p. 1.
Manchester Friends of the Earth (MFOE). (2018). 'Green Brexit? New research finds weaker environmental protections than already exist under EU law', *Manchester Friends of the Earth,* pp. 1–4.
Marlow, B. (2023a). 'Politicians are forcing electric cars on us', *The Daily Telegraph,* Saturday 1st July, p. 34.
Marlow, B. (2023b). 'Energy failure threatens what's left of manufacturing', *The Daily Telegraph,* Friday 7th July, p. 20.

Marr, A. (2023). 'Rishi Sunak, the man who isn't there', *The New Statesman,* 8th February, pp. 1–5, https://www.newstatesman.com/politics/uk-politics/2023/02/rishi-sunak-leading-european-union-brexit

Martin, D., Penna, D., & Riley-Smith, B. (2023). 'Could Sunak ally Coutinho rip up the net zero playbook?', *The Daily Telegraph,* Friday 1st September, p. 5.

Mason, C. (2023). 'Rishi Sunak seeks to grab the agenda with green gamble', *BBC News,* 20th September, pp. 1–11.

McCartney, K. (2023). 'Letters to the Editor', *The Daily Telegraph,* Sunday 9th July, p. 19.

McKay, H., & O'Leary, W. (2023). 'Rishi rows back: business anger grows over UK net zero backtracking', *E3G,* Saturday 23rd September, pp. 1–5.

McKeon, C. (2023). 'Tory launches international climate declaration to bring politicians together', *The Independent,* pp. 1–7.

Middleton, J., & Ahmed, J. (2023). 'Why Rishi Sunak's net zero bonfires will leave renters colder and poorer', *The Independent,* 23rd September, pp. 1–11.

Miliband, E. (2022). *Twitter/X,* 17th November, pp. 1–7.

Monbiot, G. (2022). 'Sunak's Britain is already broken. We need more than a general election to fix it', *The Guardian,* 26th October, pp. 1–3.

Moore, C. (2023). 'The Tories' only hope of winning is to put the ordinary voter ahead of net-zero dogma', *The Daily Telegraph,* Saturday 22nd July, p. 18.

Morning Star. (2017). 'Tories Are Polluting Your Air', Morning Star, Monday 31st July, pp. 1–2.

Murray, S. (2021). 'Postgraduates: Will the Covid recovery fuel a green jobs boom for postgraduates?', *The Guardian,* Tuesday 16th February, pp. 1–6.

Murray, J. (2022) *Twitter/X,* 7th December, p. 1.

Murray. J. (2023a). *Twitter/X,* 31st July, p. 1.

Murray, J. (2023b). 'The Tory civil war over climate is heating up – and there is no peace deal in sight', *BusinessGreen,* 3rd July, pp. 1–8, https://www.businessgreen.com/analysis/4119365/tory-civil-war-climate-policy-heating-peace-deal-sight

Murray, J. (2023c). *Twitter/X,* Tuesday 19th September, p. 1.

Murray, J. (2023d). Government confirms first wave of 21 North Sea carbon storage licenses, *Business Green,* 15th September, pp. 1–8.

Murray, J. (2023e). *Twitter/X,* Friday October 6th, p. 1.

Mustoe, H. (2023). 'Car makers "fume" at petrol engine ban "dithering"', *The Daily Telegraph,* Wednesday 26th July, p. 18.

Nelson, F. (2023a). 'Look behind Labour's reformist rhetoric and there's nothing there', *The Daily Telegraph,* Friday 7th July, p. 14.

Nelson, F. (2023b). 'Texas is boiling – and booming. Where is the climate catastrophe?', *The Daily Telegraph,* Friday 21st July, p. 16.
Nelson, F. (2023c). 'Europe is beginning to turn against the prophets of climate alarmism', *The Daily Telegraph,* Friday 19th May, p. 18.
Nelson, F. (2023d). 'Rishi Sunak has played his last, best card. Now he must pray for a miracle', *The Daily Telegraph,* Friday 1st September, p. 14.
Nelson, F. (2023e). *Twitter/X,* Wednesday 20th September, p. 1.
Nevett, J. (2023). 'Rishi Sunak denies his net zero plan is wishful thinking', *BBC News,* Thursday 21st September, p. 1.
New Statesman. (2023). 'Rishi Sunak could suffer the same fate as Jim Callaghan: In common with the former Labour Prime Minister, Sunak may find that a popular leader cannot compensate for an unpopular party', *The* New Statesman, 2nd November, pp. 1–2.
Nowak, P. (2023). *Twitter/X,* Thursday 21st September, p. 1.
Oldridge, S. (2023a). *Twitter/X,* Wednesday 20th September, p. 1.
Oldridge, S. (2023b). *Twitter/X,* Tuesday 19th September, p. 1.
Oliver, M. (2023a). 'Sunak's net zero pledge to outlaw coal power stations goes up in smoke', *The Daily Telegraph,* Sunday 25th June, p. 14.
Oliver, M. (2023b). 'Oil field that could secure UK energy supplies at risk from Labour ban', *The Daily Telegraph,* Monday 29th May, p. 4.
Oliver, M. (2023c). 'UK behind in mini-nukes race, warns Rolls-Royce', *The Daily Telegraph,* Saturday 1st July, p. 33.
Oliver, M., & Chan, S. P. (2023). 'How Britain's offshore wind industry ran out of puff', *The Daily Telegraph,* Sunday 25th June, p. 31.
Oliver, M., & Lawford, M. (2023). 'Labour's green spending threatens debt crisis', *The Daily Telegraph,* Friday 9th June, p. 16.
Orlowski, A. (2023). 'This rush to net zero risks the prosperity of the West', *The Daily Telegraph,* Monday 26th June, p. 18.
Oswick, C. (2012). Discourse analysis and discursive research. In G. Symon & C. Cassell (Eds.), *Qualitative organizational research: Core methods and current challenges* (pp. 473–491). Sage Publishing.
Oxley, J. (2023). 'Gove is right to tackle EU pollution laws blocking housing', *The Spectator,* 29th August, pp. 1–7.
Payne, S. (2023). 'Opinion: Ditching net zero would cost the Tories 1.3 million votes – it's electoral madness: Net Zero is supported by a majority of voters in every single age category in every part of England', *iNEWS,* Wednesday 26th July, p. 19, https://inews.co.uk/opinion/rishi-sunak-ditch-net-zero-climate-change-radicals-2504392

Penna, D. (2023a). 'Starmer backs critics of Khan's bigger Ulez', *The Daily Telegraph,* Friday 7th July, p. 4.
Penna, D. (2023b). 'Rees-Mogg: Dial down on net zero to win next election', *The Daily Telegraph,* Monday 24th July, p. 6.
Peston, R. (2023). *Twitter/X,* Thursday 21st September, p. 1.
Pickard-Whitehead, G. (2023). 'Europe's lurch to the far-right, and its influence on UK Conservatives', *Left Foot Forward,* 15th July, pp. 1–16, https://leftfootforward.org/2023/07/europes-lurch-to-the-far-right-and-its-influence-on-uk-conservatives/
Politics UK. (2023a) *Twitter/X,* Tuesday 19th September, p. 1.
Politics UK. (2023b). *Twitter/X,* Wednesday 20th September, p. 1.
Polman, P. (2023). *Twitter/X,* Thursday 21st September, p. 1.
Ralston, J. (2023a). *Twitter/X,* Thursday 21st September, p. 1.
Ralston, J. (2023b). *Twitter/X,* Friday 22nd September, p. 1.
Rayner, G., & Swinford, D. (2017). 'Revealed: Government's electric car revolution could cost more than £200bn', *The Daily Telegraph,* 27th July, pp. 1–3.
Razzouk, A. (2023). *Twitter/X,* Sunday 30th April, p. 1.
Renwick, D. W. S. (2009). The origins of employee well-being (EWB) in Brazil: An exploratory analysis. *Employee Relations: The International Journal, 31*(3), 312–321.
Renwick, D. W. S. (2022). 'Boris Johnson, politics and green jobs in Britain', to the *22nd European Academy of Management (EURAM) Meeting, IZHAW School of Management, Winterthur, Zurich, Switzerland,* (June), pp. 1–41.
Renwick, D. W. S. (2023) 'UK workers, political conservatism and a just transition: A reflective analysis', keynote presentation for the International Conference on "Sustainable HRM and the Future of Work", *Rennes School of Business, France,* May 16–17.
Renwick, D. W. S. (2024). 'Politics and global green jobs: a critical review and research agenda', Working Paper, Nottingham Business School, UK, pp. 1–43.
Renwick, D. W. S., Brint, A., Fiorini, P. D. C., Jabbour, A. B. J., & Jabbour, C. J. C. (2024a). Investigating voluntary workplace green behaviours in Brazil: Theory and evidence. In Y. M. Yusoff & D. W. S. Renwick (Eds.), *Green human resource management and sustainability: A view from the global south countries* (Accounting, finance, sustainability, governance and fraud book series) (1st ed., pp. 1–20). Springer Publishing. forthcoming.
Renwick, D. W. S., Clark, I., & Crichton, R. (2024b). The limitations of individual analysis and alternative framing of employee green behaviour, pp. 11–28. In Y. M. Yusoff & D. W. S. Renwick (Eds.), *(2024) green human*

resource management and sustainability: A view from the global south countries, Singapore: Springer publishing (Accounting, finance, sustainability, governance and fraud book series) (1st ed., pp. 1–483) forthcoming.
Ridley, M. (2022). 'The Tories' wind power delusion', *The Spectator,* 3rd December, pp. 1–4.
Riley-Smith, B. (2023). 'Labour's net zero plan needs 3,000 turbines', *The Daily Telegraph,* Monday 17th July, p. 1.
Riley-Smith, B., & Maidment, J. (2023). 'Ban on new petrol cars by 2030 in doubt as Sunak refuses to back policy', *The Daily Telegraph,* Tuesday 25th July, p. 1.
Riley-Smith, B., & Martin, D. (2023). 'PM urged by Cabinet to abandon eco policy', *The Daily Telegraph,* Saturday 22nd July, p. 1.
Riley-Smith, B., Gutteridge, N., & Gibbons, A. (2023). 'Oil boiler ban in doubt amid Tory backlash', *The Daily Telegraph,* Saturday 12th August, p. 4.
Savage, M. (2023). 'Sunak warned his green pledges are failing', *The Observer,* Sunday 30th July, p. 1.
Savage, M., Helm, T., & King, S. (2023a) 'Massive Labour win – followed by recrimination and self-doubt', *The Observer,* Sunday 23rd July, p. 6.
Savage, M., Helm, T., & McKie. (2023b). 'Sunak and Starmer are warned: don't dump your green pledges', *The Observer,* Sunday 23rd July, p. 1.
Shapps, G. (2023a). 'Our drive to power Britain from home will keep the lights on and bring energy bills down', *The Daily Telegraph,* Wednesday 2nd August, p. 5.
Shapps, G. (2023b). 'Labour is now the political wing of Just Stop Oil', *The Daily Telegraph,* Saturday 24th June, p. 27.
Sharma, A. (2023). *Twitter/X,* Tuesday 19th September, p. 1.
Shukman, D. (2023). *LinkedIn,* 23rd September, p. 1.
Sikka, P. (2023a). *Twitter/X,* Thursday 14th September, p. 1.
Sikka, P. (2023b). *Twitter/X,* Thursday 21st September, p. 1.
Sikka, P. (2023c). *Twitter/X,* Saturday 16th September, p. 1.
Simpson, J., & Clark, A. (2023). 'EVs cause twice the damage of petrol cars on pothole-plagued roads', *The Daily Telegraph,* Tuesday 27th June, p. 1.
Skidmore, C. (2023a). *Twitter/X,* Sunday 24th September, p. 1.
Skidmore, C. (2023b). *Twitter/X,* Tuesday 19th September, p. 1.
Sky News. (2023a). 'Nadine Dorries' resignation letter in full – as she accuses Rishi Sunak of whipping up a "public frenzy" against her', *Sky News,* Sunday 27th August, pp. 1–9.

Sky News. (2023b). 'Theresa Villiers: MP failed to declare she owned Shell shares worth £70,000 while working as environment secretary', *Sky News,* Friday 11th August, pp. 1–8, https://news.sky.com/story/theresa-villiers-mp-failed-to-declare-she-owned-shell-shares-worth-70-000-while-working-as-environment-secretary-12937653

Sky News. (2023c). 'Donald Trump says Rishi Sunak was "smart" to water down key climate pledges', *Sky News,* Sunday 24th September, pp. 1–5.

Slawson, N., & Gecsoyler, S. (2023). 'Two-thirds of people do not trust Tories on environment, poll finds – as it happened', *The Guardian,* Friday 22nd September, pp. 1–14.

Smyth, C. (2023a). 'Tory retreat from green policies to woo voters', *The Times,* Monday 24th July, p. 1.

Smyth, C. (2023b). 'Sunak sticking to 2030 ban on sales of new petrol cars', *The Times,* Wednesday 26th July, p. 10.

Somerville, E. (2023). 'Extortionate' heat pump scheme will 'take us back to the Dark Ages', *The Daily Telegraph,* Saturday 12th August, p. 4.

Stanley, T. (2023). 'Britain has become a world leader in net zero fantasies and delusions', *The Daily Telegraph,* Monday 31st July, p. 16.

Starmer, K. (2023). 'News', *The Observer,* 16th July, p. 5.

Stone, S. (2023). 'Climate policy rollback risks Tory seats: Poll reveals "climate majority" in Blue Wall and key marginals', *Business Green,* 23rd September, pp. 1–9.

Stop Cambo. (2023). *Twitter/X,* Thursday 14th September, p. 1.

Sun, P., Doh, J., Rajwani, T., Werner, T., & Luo, X. R. (2023). 'The Management of Socio-Political Issues and Environments: Toward a Research Agenda for Corporate Socio-Political Engagement', *Journal of Management Studies,* First published 21st September 2023, https://doi.org/10.1111/joms.13002.

The Daily Telegraph. (2023a). 'Letters to the Editor: The PM has put common sense before ideology – but will have to face down the net zero zealots', *The Daily Telegraph,* Thursday 21st September, p. 15.

The Daily Telegraph. (2023b). 'Conservatism's future', *The Daily Telegraph,* Wednesday 17th May, p. 15.

The Daily Telegraph. (2023c). 'Just stop Labour's ban', *The Daily Telegraph,* Monday 29th May, p. 15.

The Daily Telegraph. (2023d). 'Tories must stop tearing themselves apart', *The Daily Telegraph,* Sunday 11th June, p. 19.

The Daily Telegraph. (2023e). 'Labour's energy plan', *The Daily Telegraph,* Monday 19th June, p. 15.

The Daily Telegraph. (2023f). 'Time to reconsider our approach to net zero', *The Daily Telegraph,* Saturday 24th June, p. 19.
The Daily Telegraph. (2023g). 'The net-zero poll tax', *The Daily Telegraph,* Saturday 24th June, p. 17.
The Daily Telegraph. (2023h). 'The anti-green revolt is the Tories big chance', *The Daily Telegraph,* Saturday 22nd July, p. 17.
The Daily Telegraph. (2023i). *The Daily Telegraph,* Tuesday 25th July, p. 15.
The Daily Telegraph. (2023j). 'Sunak strikes back against anti-car zealots', *The Daily Telegraph,* Sunday 30th July, p. 19.
The Daily Telegraph. (2023k). 'The truth about net zero', *The Daily Telegraph,* Saturday 12th August, p. 17.
The Daily Telegraph. (2023l). 'Sunak is finally standing up to the green Blob', *The Daily Telegraph,* Sunday 3rd September, p. 17.
The Daily Telegraph. (2023m). 'Britain needs a referendum on net zero', *The Daily Telegraph,* Sunday 23rd July, p. 17.
The Daily Telegraph. (2023n). 'Tory women jostle for position as next in line', *The Daily Telegraph,* Tuesday 3rd October, p. 6.
The Economist. (2023). 'Meet Ms Heeves, the face of Britain's new political consensus: How Labour and the Conservatives ended up agreeing on almost everything', *The Economist,* 4th February, p. 26.
The Guardian. (2023a). 'Britain's electoral map is being radically redrawn by the voters' themselves', *The Guardian,* Friday 21st July, p. 2.
The Guardian. (2023b). 'Editorial: The Guardian view on diluting net zero targets: bad economics dictated by cynical politics', *The Guardian,* Wednesday 20th September, pp. 1–5.
The London Economic. (2023). 'Don't Look Up? Fury as GB News hosts "in hysterics" over climate change warnings', *The London Economic,* 20th August, https://www.thelondoneconomic.com/news/gb-news-dont-look-u-climate-change-interview-laughing-354968/
The Sunday Times. (2023). 'Labour must show voters a comprehensive vision', *The Sunday Times,* 23rd July, p. 24.
The Times. (2023b). *LinkedIn,* 23rd September, p. 1.
The Times. (2023c). 'Times letters: Rishi Sunak's green gamble over net zero strategy', *The Times,* Saturday 23rd September, p. 1.
Thomas-Peter, H. (2023). 'Government progress on net zero "worryingly slow", report says', *Sky News,* Wednesday 28th June, pp. 1–7, https://news.sky.com/story/government-progress-on-net-zero-worryingly-slow-report-says-12910771#:~:text=The%20Climate%20Change%20Committee%20says,it%20was%20a%20year%20ago

Timothy, N. (2023). 'The problems Britain faces are huge, yet our politics has never felt so small', *The Daily Telegraph,* Monday 19th June, p. 14.
Tominey, C. (2023a). 'It's time Rishi Sunak stood up for the silent majority driven to despair by eco-fanaticism', *The Daily Telegraph,* Saturday 1st July, p. 21.
Tominey, C. (2023b). 'Net zero extremism is turning into a middle-class plot against the poor', *The Daily Telegraph,* Saturday 29th July, p. 19.
Tosi, J., & Warmke, B. (2020). *Grandstanding: The use and abuse of moral power* (pp. 1–226). Oxford University Press.
Toynbee, P. (2023). 'Opinion, Labour: The Tories have laid a "cut the green crap" trap for Keir Starmer. He must not fall for it', *The Guardian (Journal),* Tuesday 25th July, pp. 2–3, https://www.theguardian.com/commentisfree/2023/jul/25/tories-cut-green-trap-keir-starmer-policies-bills-labour
Valero, A, Li, J., Muller, S., Riom, C., Nguyen-Tien, V., & Draca, M. (2021) 'Policy brief: Are "green" jobs good jobs? Lessons from the UK and EU to inform labour market transitions', Grantham Research Institute on Climate Change and the Environment/Centre for Climate Change Economics and Policy, *London School of Economics and Political Science,* October, pp. 1–8, https://www.lse.ac.uk/granthaminstitute/publication/are-green-jobs-good-jobs-how-lessons-from-the-experience-to-date-can-inform-labour-market-transitions-of-the-future/
Vaughan, A. (2023a). '"Red" Tories: Sunak isn't green enough', *The Times,* Wednesday 26th July, p. 8.
Vaughan, A. (2023b). *Twitter/X,* Wednesday 20th September, p. 1.
Vince, D. (2023). *Twitter/X,* Thursday 21st September, p. 1.
Vinter, R., & Stacey, K. (2023). '"We need more investment, not less": UK voters on Sunak's net zero rollback', *The Guardian,* Wednesday 20th September, pp. 1–3.
Walker, J. (2023a). 'Rishi Sunak's family firm Infosys signed $1.5B deal with BP', *The National,* 2nd August, pp. 1–7, https://www.thenational.scot/news/23697572.rishi-sunaks-family-firm-infosys-signed-1-5b-deal-bp/
Walker, B. (2023b). 'The fall of Brand Sunak', *The New Statesman,* 22nd September, pp. 1–5.
Walker, C. (2023c). *Twitter/X,* Wednesday 20th September, p. 1.
Walker, C. (2023d). *Twitter/X,* Saturday 23rd September, p. 1.
Warner, J. (2023a). 'The sums don't add up for offshore wind as an effective route to net zero', *The Daily Telegraph,* Wednesday 30th August, p. 21.
Warner, J. (2023b). 'Labour leaders will regret making net zero such a key election issue', *The Daily Telegraph,* Sunday 11th June, p. 26.

Wells, I. (2023). 'Rishi Sunak scraps home energy efficiency taskforce', *BBC News,* 23rd September, pp. 1–10.
Wheeler, C., & Dahan, M. (2023). 'I hate tree-huggers – Kier Starmer explodes over green policy', *The Sunday Times,* Sunday 9th July 2023, pp. 1–4, https://www.thetimes.co.uk/article/i-hate-tree-huggers-keir-starmer-explodes-over-green-policy-6hhnj9r9x
Wheeler, C., Boothman, J., & Spencer, B. (2023) 'Sunak to pour millions into North Sea oil', *The Sunday Times,* 30th July, pp. 1–2.
Wilkinson, A., Knoll, M., Mowbray, P. K., & Dundon, T. (2021). New trajectories in worker voice: Integrating and applying contemporary challenges in the Organization of Work. *British Journal of Management, 32*(3), 693–707. https://onlinelibrary.wiley.com/doi/10.1111/1467-8551.12528
YouGov. (2023). *Twitter/X,* Friday 22nd September, p. 1.
Yusoff, Y. M., & Renwick, D. W. S. (2024). *Green human resource management and sustainability: A view from the global south countries* (Accounting, finance, sustainability, governance and fraud book series) (1st ed., pp. 1–483). Springer Publishing. forthcoming.

Index[1]

[1] Note: Page numbers followed by 'n' refer to notes.

D. W. S. Renwick, *Work, Politics and the Green Industrial Revolution*,
https://doi.org/10.1007/978-3-031-53067-8

www.ingramcontent.com/pod-product-compliance
Ingram Content Group UK Ltd.
Pitfield, Milton Keynes, MK11 3LW, UK
UKHW020836290926
15470UKWH00004B/1387
9783031530661